Oil and Nationalism in Argentina

CARL E. SOLBERG

Oil and Nationalism
in Argentina
A History

Stanford University Press, Stanford, California
1979

Stanford University Press, Stanford, California
© 1979 by the Board of Trustees of the
Leland Stanford Junior University
Printed in the United States of America
ISBN 0-8047-0985-8   LC 77-92947

Published with the assistance of the
  Andrew W. Mellon Foundation

Illustrations 13 and 20 appear with
permission of Yacimientos Petrolíferos Fiscales;
all others appear with permission of the
Archivo General de la Nación, Buenos Aires.

For DORIAN GRAHAM MUNCEY

Preface

The rise of nationalistic petroleum policies in the underdeveloped countries is one of the most significant political phenomena of our times. Convinced that rapid, sustained economic growth requires national ownership of oil resources, a growing number of countries in Latin America, as elsewhere in the underdeveloped world, are challenging the international distribution of power and wealth. Although Venezuela's recent decision to nationalize its oil industry has attracted much attention in the United States, it represents only the latest step in a movement of Latin American petroleum nationalism that dates back well over half a century. That movement originated in Argentina, the first country outside the Soviet Union to form a vertically integrated state-owned petroleum industry. Hence, the experience of the Argentine state oil company, *Yacimientos Petrolíferos Fiscales* (YPF), provides a suggestive case study of petroleum development in the context of economic nationalism.

Today the industrializing Argentine economy depends heavily on the republic's oil resources, and Argentina is one of South America's principal petroleum-producing countries. Argentina's major oil resources lie in three regions. The northernmost runs for about 450 kilometers along the densely wooded and sparsely populated eastern foothills of the Andes in the provinces of Salta and Jujuy; the oil-bearing formation extends north across the frontier into southeastern Bolivia. The petroleum from this region is of high quality, but exploration and drilling are difficult and costly because of the rugged terrain. Argentina's second oil region, which today produces about a third of the country's total output, begins about 900 kilometers south of the lower end of the northern region. It, too, parallels the Andes, and extends almost 900 kilometers from a point about 60 kilometers northwest of the city of Mendoza to the vicinity of the Río Negro. The arid and isolated high plains in this region, particularly in Neuquén, long impeded development of its oil resources. The southernmost of the three principal oil regions, the Patagonian coast in Santa Cruz and Chubut provinces, was long the most famous and productive oil formation in Argentina. Centered at the Atlantic coast city of Comodoro Rivadavia, the fields extend in an arc with a radius of about 150 kilometers. Comodoro Rivadavia's fields no longer supply the majority of the republic's petroleum, as they did for decades, but new fields in Santa Cruz province now produce large amounts, and promising strikes have been made in the island territory of Tierra del Fuego.

But these resources remained largely untapped until 1907, when oil was discovered at Comodoro Rivadavia. Prior to that year, and for many years thereafter, Argentina relied almost totally on coal for fuel. Along with the bulk of the republic's manufactured goods, the coal was imported, chiefly from Great Britain. And the agricultural products of the pampas were exported to pay for these imports, chiefly to Great Britain. The First World War upset this traditional system, however, and the resulting crisis led some prominent figures to challenge the basic premises of Argentina's export-oriented economy. Increasingly, national leaders became convinced that the republic's future prosperity required greater self-sufficiency. These precursors of modern Argentine economic nationalism argued that the republic's dependence on foreign capital, the export trade, and imported fuel left Argentina virtually a pawn in the hands of foreign powers.

The early leaders of YPF, who shared these visions of greater economic independence for Argentina, developed a theme of economic nationalism that this book calls petroleum nationalism. Briefly, the petroleum nationalists emphasized that Argentina's oil resources, if developed by a state monopoly, would enable the country to achieve sustained economic development without interference from the international oil companies or their home governments. By the late 1920's, petroleum nationalism not only had attracted powerful political support but also had become a major ideological force. The emergence of petroleum nationalism within the context of Argentine political and economic history is the subject of this book.

The relationships among ideology, politics, and socioeconomic structures examined in this study pose some basic questions about twentieth-century Argentine development. Which social and economic groups supported petroleum nationalism, and why? How did political parties and leaders attempt to mobilize this support? What was the role of the military in YPF's development? How did regionalism and federalism affect the rise of petroleum nationalism? What was the impact of petroleum nationalism on government policy formulation? What are the results of the use of petroleum nationalism as a political tool in a dependent economy? Did Argentina's transition from an elite-dominated to a mass-participatory society give further impetus to nationalism in the oil sector?

The book considers each of these problems through historical analysis of Argentine petroleum politics and YPF's development from 1907 to the mid-1970's. Attention, however, focuses on the crucial formative decades of the state oil industry, a period that coincided with Argentina's liberal political experiment between 1916 and 1930. It was during these years that Enrique Mosconi, the guiding genius of YPF, and Hipólito Yrigoyen, the predominant figure in early-twentieth-century Argentine politics, established the institutions and ideologies that have shaped Argentine petroleum affairs during the remainder of the century.

In this study I have used the metric system to express all petroleum measurements. The basic unit is the cubic meter (m^3). Historically, the U.S. oil trade has used the barrel (42 U.S. gallons) and the British the ton (the unit of the shipping trade); but since the Argentine government has always employed the metric system to measure both domestic production and imports, I have standardized all the tables and the text to conform to Argentine practice. For conversion purposes, one cubic meter of crude oil equals 6.3 barrels or .93 metric ton.

On monetary matters, too, I have followed standard Argentine practice, which is to quote international financial transactions in gold pesos and domestic transactions in paper pesos. The value of the gold peso from 1899 to 1933 was U.S. $.9658. During periods of convertibility, the legal value of paper pesos circulating inside Argentina was .44 of the gold peso.

Many organizations and individuals have aided the preparation of this book. The University of Washington's Graduate School Research Fund granted financial support that permitted employment of a capable research assistant, E. Louise Flippin, during the preliminary stage of work, and the same source extended support for the final preparation of the manuscript. The Social Science Research Council, which supported the research with a Post-Doctoral Fellowship in 1973 and 1974, made possible an extended period of study in Argentina. In Buenos Aires, the superb collections at the *Biblioteca Tornquist* proved invaluable for the history of YPF. The staffs of the *Biblioteca Nacional* and of the *Biblioteca del Instituto Torcuato di Tella* cooperated generously. Pat Léger, of the Simon Fraser University Library in Vancouver, British Columbia, extended valuable help in locating and obtaining relevant British Foreign Office diplomatic correspondence. Ruth Kirk and her staff at the Interlibrary Loan Office of the University of Washington Library also deserve praise for their patient and capable assistance. Dauril Alden, Stanley Payne, Joseph Tulchin, John Wirth, Stephen J. Randall, Daniel Greenberg, David Tamarin, and my colleagues in the History Research Group at the University of Washington read and criticized the manuscript at various stages. Finally, my special thanks go to Brigid King for her help in preparing and typing the manuscript.

C.E.S.

Contents

Illustrations

Tables

Oil and Nationalism in Argentina

BOLIVIA

· Santa Cruz

Camiri

BRAZIL

Yacuiba
JUJUY ① · Tartagal
· Orán
② · Jujuy

PARAGUAY

CHILE

Salta ·
SALTA

FORMOSA · Asunción

TUCUMÁN

CHACO

SANTIAGO
DEL
ESTERO

CATAMARCA

MISIONES

Río Uruguay

SANTA FE

LA RIOJA

CÓRDOBA

Santa
Fe

CORRIENTES

SAN JUAN

Córdoba

ENTRE
RÍOS

Mendoza ·
③

SAN
LUIS

Rosario ·

URUGUAY

Zárate ·

MENDOZA

Buenos Aires ·
La Plata

Montevideo ·

Río de la Plata

BUENOS AIRES

LA PAMPA

Mar del Plata ·

NEUQUÉN

Río Colorado

Bahía Blanca ·

Plaza Huincul /
④ · Neuquén

Río Negro

RÍO NEGRO

0   250   500
KILOMETERS

Río Chubut

CHUBUT

PRINCIPAL ARGENTINE OIL FIELDS

Colonia
Sarmiento
Comodoro Rivadavia ·
⑤ GOLFO DE
SAN JORGE

1 Orán-Tartagal   4 Neuquén
2 Jujuy   5 Comodoro Rivadavia
3 Mendoza

Stanley

ISLAS MALVINAS
(FALKLAND ISLANDS)

SANTA CRUZ

TIERRA DEL FUEGO

Argentina, Showing Provinces and Principal Oil Fields

CHAPTER I

The Struggle for State-Owned Oil, 1907-1914

Petroleum altered the course of Argentine political and economic history, but when government drillers struck oil on state land in Patagonia in 1907, the discovery stirred little excitement. The major newspapers in Buenos Aires hardly mentioned the event—an ironic attitude in view of the fact that South America's second-largest country depended almost entirely on imported fuel. Yet Argentina was in the midst of a major export boom, and the powerful landed elite was far more concerned with shipping ever greater quantities of meat and cereals to Europe than with developing an oil field on the remote Patagonian frontier. Free-trade liberalism dominated Argentine economic thought, and it is unlikely that the state would have become involved in the petroleum business as early as it did had oil not been found on government land. Though much of the political elite viewed the state's new oil field with apathy or even with hostility as an infringement on the ideals of laissez-faire capitalism, a few highly placed government officials fought to maintain state ownership and to increase production and extend exploration. This chapter focuses on their ideas and their accomplishments, which laid the foundation for Argentine petroleum nationalism in the 1920's.

The Economic Setting

A brief survey of Argentina's economic structure in the early years of the twentieth century will provide the necessary background for an analysis

1

of the origins and development of the state petroleum industry. Argentine foreign trade flourished, but the country was essentially an economic colony of Great Britain. Firmly tied to European markets and reliant on foreign, primarily British, capital investment, the economic structure of Argentina was geared toward one basic activity: the production and export of huge quantities of agricultural and cattle products.

When the export boom reached its height in 1913, about eight million people inhabited Argentina's area of slightly over 2.5 million square kilometers stretching from the subtropics bordering Paraguay and Bolivia to the fringes of the Antarctic 3,700 kilometers away. Then as now the population was concentrated in the pampas, immense grasslands that contain some of the world's richest soil and that fan out for some 500–600 kilometers from Buenos Aires, the capital. Almost uncultivated as late as 1870, the pampas were subsequently transformed in response to the rapidly rising demand for cereals and meat in Europe. As the export boom accelerated, a few thousand large landowners, who enjoyed near-monopoly ownership of the country's best farming and cattle-raising land, profited enormously. This landed class consolidated its position by continuing to buy up the increasingly valuable pampas farther from Buenos Aires and bringing them into production. The effect of this development was an expansion of the area under cultivation from about 500,000 hectares in 1870 to 24.1 million hectares by 1913. To work the land, a huge new labor force was needed, and into this labor vacuum swept a flood of poor Europeans. Between 1870 and 1914, about 3.5 million immigrants, primarily Italians and Spaniards, settled in Argentina. Most became tenant farmers or sharecroppers, but many of the newcomers, frustrated by the difficulty of actually acquiring land, remained in or returned to the cities, particularly Buenos Aires, where they formed a large segment both of the urban workers and of the nascent middle classes.[1]

The transformation brought about by immigrant labor and agricultural expansion led to a spectacular growth in Argentine exports—from a value of 22 million gold pesos in 1862 to a value of 519 million in 1913. At least 30 percent of national production was shipped abroad each year; by 1927, when exports absorbed 42 percent of the country's production, Argentine agricultural exports were second only to those of the United States.[2] Argentina had become the world's leading exporter of linseed and corn, the third or fourth largest exporter of wheat, and by far the prime source of beef for the British market. These products, along with one or two other staples, accounted for at least 90 percent of the republic's foreign exchange earnings.[3]

To function efficiently, the export trade required a modern transport system to carry agricultural products to exporting points and to distribute imported goods throughout the country, and by the First World War Argentina had acquired one of the world's most extensive railway networks. The development of this network—which boasted 33,219 kilometers of track in 1913, one of the world's larger railway systems—required immense capital investments that totaled 1.4 billion gold pesos by 1914. British companies, which had invested over one billion gold pesos in Argentine railways, controlled the lion's share of the republic's rail network. The remainder was owned by French concerns and by the Argentine State Railways, which had built a series of feeder and connecting lines, primarily in remote provinces.[4] However, about 65 percent of all track was located in the three principal pampas provinces of Buenos Aires, Córdoba, and Santa Fe; this meant that most provinces outside the pampas, as well as the more remote regions, lacked adequate transport facilities. The diversity of gauges—there were three major ones—further impeded integrated economic development.[5]

The railways were only the most visible symbol of the predominant role foreign capital had attained in the prewar Argentine economy. By 1913, foreign investments reached a total of 3.25 billion gold pesos and financed about 50 percent of Argentina's fixed capital assets. Foreigners owned most of the utilities and communications networks, the major port facilities, the largest private banks, scores of large commercial enterprises, most of the meat-packing plants, and many of the big cattle and sheep ranches. Service charges on foreign investments absorbed about 35 percent of Argentina's total export earnings each year. In addition to the direct investments, foreigners held much of the Argentine public debt and an estimated 500 million gold pesos in private mortgages (in 1918). As they were with the railways, the British were the leading investors in most other enterprises. In 1913, total British investments in Argentina reached 1.928 billion gold pesos, compared to French investments of 475 million and German investments of 250 million. (By contrast, U.S. holdings amounted to only 40 million.)[6] The dominant position Great Britain enjoyed in the prewar Argentine economy extended far beyond its investments, however: the bulk of Argentine trade moved in British ships; and Great Britain consistently purchased more of Argentina's exports, and supplied more of her imports, than did any other country. As E. J. Hobsbawm has put it, Britain's economic power made Argentina a "British informal colony," and an "honorary dominion." [7]

Few manufacturing industries were able to prosper in this export economy, for nineteenth-century Argentine governments, dominated by the rural elite, had kept tariffs low and had designed them to produce revenue rather than stimulate industrialization. Most of the factories that did exist —notably meat-packing plants—processed or refined rural products for export. Industries that produced primarily for the national market achieved significant tariff protection only in exceptional cases—the wine industry of Mendoza province and the sugar industry of Tucumán being prime examples. Although manufacturing of some basic consumer goods grew during the First World War and in the immediate postwar years, the industrial sector remained on the whole small and weak, and imported goods continued to provision the Argentine market. [8] By way of example, as late as 1929 imports supplied 92 percent of the republic's total consumption of cotton textiles. [9]

Fascinated by the export boom, the massive immigration, the formation of a large urban middle class, and the glittering façade of material prosperity in the great metropolis of Buenos Aires, contemporary observers convinced themselves that Argentina was rapidly emerging as a modern, developed nation. Already Latin America's most prosperous country by 1913, Argentina's Gross Domestic Product per capita in that year exceeded those of France, Italy, and Sweden. [10] But such comparisons are misleading, for Argentine prosperity was built on shaky foundations that were defenseless against changing international political and economic circumstances. Moreover, whole regions and social classes never enjoyed the benefits of the export boom, which were concentrated primarily in the capital and the pampas region. Unable to compete with European imported goods, the domestic industries of several northern and western provinces long had been in a state of decline that left the inhabitants of much of the interior in increasing poverty. Plagued by high rates of unemployment, illegitimacy, and disease, the Argentine northwestern provinces resembled the underdeveloped regions of neighboring South American countries more than the flourishing pampas region or the capital. [11]

The superficial opulence of Buenos Aires, a city of 1.5 million people in 1914, concealed the bleak living standards that much of the urban working class endured. Prices of most products in Buenos Aires were considerably higher than in Western Europe, a fact often overlooked by analysts impressed by the country's aggregate growth during this period. In 1910, a loaf of bread cost more in the capital of wheat-exporting Argentina than in London, Paris, or New York. Contemporary observers bitterly criticized the high prices of consumption goods: "Everything is expensive. A man needs three times the salary in Buenos Aires to live the same way he would live in London," complained one British writer in 1913.[12] Housing was notoriously scarce and dear. "The price of housing was . . . so high that the only manner of meeting it was to reduce the habitation to the minimum size, to one room. . . . Eighty of every hundred families occupied a single room," wrote economist Alejandro E. Bunge.[13]

Near-total reliance on imported fuel was another structural weakness of the Argentine economy during the export boom. Copious supplies of wood and a trickle of oil made up the republic's entire fuel production as late as 1913. The wood, produced in the heavily forested northern provinces and used primarily for locomotive fuel, supplied only about 15 percent of Argentina's energy and yielded far less energy per ton than did coal, which supplied the bulk of the country's energy.[14] And for decades it had been the coalfields of South Wales that had fueled the Argentine economy. Favored by the British shipping industry, which offered low freight rates for coal on otherwise partly empty ships bound for Argentina, Welsh coal enjoyed advantages other coal producers could not match and provided 90 percent of Argentina's coal imports.[15] Moreover, the large British-owned coal-consuming enterprises—particularly the railways, which consumed over 70 percent of the coal imports—relied on British coal as a matter of policy.[16] Largely because of the railways' thriving demand, coal imports grew enormously; by the eve of the First World War, Argentine consumption of Welsh coal was four times what it had been only a decade earlier.

Despite recognition of the dangers of reliance on imported coal, before the First World War neither the government nor private investors directed much attention to developing Argentina's own coal resources. During the 1870's and 1880's, prospectors made promising discoveries in Mendoza and San Juan provinces and in Patagonia, but no one had any real idea of the extent of Argentina's coal resources until 1902, when Wenceslao Escalante, Minister of Agriculture under President Julio Roca (1898–1904), commissioned Enrique Hermitte, a government mining engineer, to survey and test national coal deposits. Impressed by the high quality of much of the coal he examined, Hermitte reported that some was as good as

Welsh coal and urged the state to promote mining development.[17] President Roca took a small step in that direction in 1904 by creating a Bureau of Mines, Geology, and Water Resources in the Ministry of Agriculture and appointing Hermitte to head it; however, the new agency suffered from lack of funds and was unable to map the republic's mineral resources adequately.[18] The government, assured a steady supply of cheap imported coal, took no further action to develop Argentine coal resources.*

Petroleum, still something of a novelty among Argentine consumers, supplied less than 5 percent of Argentina's energy on the eve of the First World War. In the absence of significant internal production, petroleum products were imported by a few foreign-owned companies, the largest of which was The West India Oil Company (WICO). A subsidiary of Standard Oil of New Jersey, WICO had been organized in 1902 to market petroleum products throughout Latin America. In 1911, WICO, already a large importer of fuel oil, went into the refinery business in Argentina when it purchased control of the Compañía Nacional de Petróleos, Limitada (CNP), an Argentine firm. Organized in 1905, CNP had built a small refinery at Campana, near Buenos Aires, to produce kerosene and gasoline behind the protection of a tariff wall. WICO made major investments in its new subsidiary—which continued to operate under its own name, though Argentine participation in it henceforth was limited to 35 percent of the stock—and greatly expanded the Campana refinery's capacity.[19] As a result, by 1917 WICO (through CNP) supplied 95 percent of Argentina's kerosene and 80 percent of its gasoline. Its commanding position in the market, and its high profits (which between 1912 and 1915 averaged 21.2 percent of invested capital annually[20]), brought Jersey Standard under severe Argentine criticism during the First World War, as we shall see in the next chapter.

The Anglo-Mexican Petroleum Products Company Limited, a subsidiary of Royal Dutch Shell, became Jersey Standard's principal competitor in importing petroleum products into Argentina. Anglo-Mexican, which marketed products of Shell's Mexican Eagle Oil Company, did not open its first office in Buenos Aires until 1913, but prospered rapidly after the outbreak of war disrupted coal shipments. By 1915, it was importing over

---

*Indeed, systematic exploration of coal deposits did not begin until the Second World War, when detailed study of the Río Turbio fields, originally discovered in 1887 and located 260 kilometers west of Río Gallegos in Santa Cruz, began. The estimated reserves there are 450 million tons of subbituminous coal. Production, which started in 1946, reached about 625,000 tons annually by the mid-1970's.

10,000 cubic meters of Mexican oil per month. To serve its major customers, which included British meat-packing plants and railways, Anglo-Mexican built major storage facilities at Buenos Aires and Bahía Blanca. Though it competed with WICO as a major importer of fuel oil, Anglo-Mexican did not deal in kerosene or gasoline until after the war.[21]

Early Exploration for Oil in Argentina and the 1907 Discovery

Before the discovery of oil in Patagonia in 1907, the Argentine government displayed the same lack of concern about the location and development of petroleum deposits that it did about coal resources. The existence of petroleum in Argentina had been confirmed as early as the mid-seventeenth century, when a Franciscan friar described visible oil pools in Salta, a province in the northern Andean foothills near Argentina's present borders with Chile and Bolivia; but until the twentieth century the republic's petroleum resources attracted little more than occasional scientific curiosity. The First National Census of 1869, which reported important oil deposits 70 leagues (some 350 kilometers) south of Mendoza, an Andean foothill city almost due west of Buenos Aires, failed to spark the interest of either the government or private investors.[22] During the 1870's, sporadic private explorations were made in Salta and Jujuy provinces, but these projects soon failed. The sole success story of Argentine oil production before the 1907 Patagonian strike began when the Compañía Mendocina de Petróleo drilled 20 wells, four of which proved productive, near Mendoza in 1887. The company built a small pipeline to ship petroleum to Mendoza, and produced some 8,600 cubic meters of oil between 1887 and 1891. It abandoned operations in 1897, though, after losing money in an unsuccessful venture in Jujuy and having its shallow Mendoza wells run dry.[23]

For a decade after the Mendoza company failed, Argentina produced no oil, and no significant exploration took place. In fact, when Patagonian oil finally was discovered in 1907, a search for water, not petroleum, was the ostensible reason for the drilling expedition. Comodoro Rivadavia, the site of the search, was located on a bleak fringe of Argentine civilization. A coastal village of 50 permanent families founded in 1899, Comodoro Rivadavia had an economy centered on the export of wool, the main product of Chubut, then a national territory. Its port was primitive and unprotected by breakwaters, and a narrow-gauge state-owned railway, still under construction in 1907, reached inland only as far as agricultural settlements at Colonia Sarmiento, 120 kilometers away.[24] The environment of this part of Patagonia was desolate. Seemingly endless flatlands stretched to the horizon, broken only by occasional dry valleys and deep canyons. The dry climate and the strong winds that constantly swept the region stunted all vegetation and led to numerous cases of deafness and some of madness among the inhabitants. Visiting agronomist Hugo Miatello described these "damned winds" as the "true calamity of Patagonia," the "nightmare both of the populace and of visitors."[25]

Water was exceedingly scarce and had to be brought in to Comodoro Rivadavia, where it was sold at high prices. The nearest source, undependable and polluted, was five kilometers from the town, but the major supply came from an oasis 50 kilometers inland known as Behr's Estancia, where a group of Boer immigrants had settled in 1903 and carried on almost the only agriculture in the district. The Boers were willing to ship water by train to Comodoro Rivadavia, but the supply was often intermittent in summer, when they irrigated their alfalfa crop. Faced with the constant specter of drought, Comodoro Rivadavia pleaded to the national government to search for water.[26]

In response, the Bureau of Mines, Geology, and Water Resources sent a drilling team, which spent much of 1907 in an unsuccessful search. Some Argentine historians argue that the government was really looking for oil rather than water, an allegation strengthened by Hermitte's disclosure in 1904 that the Minister of Agriculture had ordered the Bureau to explore not only for coal and water but also for petroleum. Scientists long had suspected that the Patagonian coast, which contains widespread outcroppings of asphalt, was likely oil territory, and an unsuccessful water search at Comodoro Rivadavia in 1903 had discovered traces of gas. In any case, Julio Krause, who headed the Bureau's Water Resources Department, hired European oil-drilling experts to work on the 1907 Comodoro Rivadavia drilling project.[27]

On December 13, 1907, the drillers struck oil. The villagers, whose hopes for water were temporarily dashed, were not particularly impressed and were on the point of expelling the drillers, who were jubilant as the oil gushed up the wooden drilling tower.[28] José Fuchs, chief of the drilling party, rushed to telegraph president José Figueroa Alcorta, who responded the next day with an executive decree creating a national oil reserve, on which private concessions were prohibited, on about 200,000 hectares of land five leagues (some 25 kilometers) in all directions from Comodoro Rivadavia. The discovery was made on government-owned land, and hence

1. Oil drilling at Comodoro Rivadavia, 1907. This was the site of the December 13 discovery.

the president based his decree on the Public Lands Law of 1903, which permitted the chief executive to create reserves when petroleum was discovered on state-owned lands. Figueroa Alcorta's decree established the precedent for the sort of state oil development characteristic of modern Argentina. Yet the President hesitated to promote production on the state reserve, emphasizing instead the need to gather information on the extent and quality of the deposits first. Accordingly, the Ministry of Agriculture ordered the Bureau of Mines to continue exploratory drilling, to conduct tests of the petroleum's quality, and to store as best it could whatever oil was produced. But the underfinanced Bureau was able to work only very slowly.[29]

A small team of workers at Comodoro Rivadavia struggled with the problems of organizing petroleum production. Lacking storage tanks and piping, the workers dug a 5,000-cubic-meter storage area in the ground to receive the output of the first wells. After acquiring a second drilling machine, the crews sank four more wells by 1910, three of which yielded oil. But production remained primitive, grew slowly, and totaled only

TABLE I.I. *Argentine Domestic Production and Imports of Petroleum,*
1907–14
(Cubic meters)

Year	Total petroleum consumption	Domestic production	Percent of total consumption	Imports	Percent of total consumption
1907	196,929	16	—	196,913	100.0%
1908	125,828	1,821	1.4%	124,000	98.6
1909	147,989	2,989	2.1	145,000	97.9
1910	165,293	3,293	2.0	162,000	98.0
1911	172,082	2,082	1.2	170,000	98.8
1912	211,462	7,462	3.5	204,000	96.5
1913	300,733	20,733	6.9	280,000	93.1
1914	278,795	43,795	15.7	235,000	84.3

SOURCES: Dorfman, p. 143; "Resumen estadístico de la economía argentina," *Revista de Economía Argentina*, 20 (Nov. 1938), p. 323.

8,110 cubic meters by the end of 1910 (see Table 1.1). The 5,000-cubic-meter open-air storage facility was apparently never overtaxed, since more than a third of the petroleum stored in it seeped out or evaporated. The remainder was eventually shipped to Buenos Aires for tests or consumed locally, primarily by the Colonia Sarmiento railway and an electric power plant.[30]

Political Crisis and Oil-Policy Drift

Why did the national government respond so slowly to the needs of the new state oil fields? The growing Argentine political crisis between 1907 and 1912 provides part of the explanation, for demands for political reform overshadowed all other issues and occupied most of the government's attention. Public indignation was rising against the openly rigged elections, particularly common in the interior, that excluded the masses and the middle classes from effective political participation and left the only parties in the national Congress before 1912 (with very few exceptions) a handful of personalistic factions responsive to various segments of the elite. In the years between 1907 and 1912, increasingly discredited by their use of electoral fraud, these traditional conservative parties were disintegrating.

The largest remained the Partido Autonomista Nacional (PAN), a coalition of provincial conservative parties historically loyal to General Julio Roca, Argentina's late-nineteenth-century military hero and political strongman (President 1880–86, 1898–1904). In power, the PAN faithfully adhered to Roca's dictum "peace and administration," and to his free-trade, laissez-faire economic policies. A second major grouping calling itself the Partido Autonomista usually dominated elections in the capital and followed the influential ex-President Carlos Pellegrini (Vice-President 1886–90; President 1890–92). A number of smaller parties continued to represent the political machines of various conservative provincial governors.[31] But these "parties," whose leadership, recruitment, and policies traditionally had ignored the masses and the middle classes, no longer enjoyed undisputed control over Argentine politics. Since the mid-1890's, a small but vigorous Socialist Party had been working diligently to establish an electoral base among the discontented working classes of the capital. The party scored its first victory when it elected a candidate to the national Chamber of Deputies in 1904. The Socialists remained a relatively small group, however, for their emphasis on participation in elections and the achievement of social change by political means failed to appeal to the anarchist and syndicalist trade unions that controlled the bulk of the capital's labor movement. Dedicated to the use of the strike as a tactic for change, the anarchist and syndicalist unions grew increasingly militant after 1900. The several general strikes they mounted deeply alarmed the upper class and stimulated elite politicians to search for some sort of political reform to dispel the specter of mass rebellion.

Reformists in the political elite began to seek an accommodation with another emerging force in Argentine politics, the Unión Cívica Radical, or Radical Party, a coalition of members of the middle classes and newly rich landowners that had been organizing nationwide since the early 1890's. The Radicals concentrated their appeal on the rapidly growing middle classes, which constituted about 30 percent of the country's population by 1914. With their aspirations for upward mobility stimulated by the rapid expansion of the educational system and by the booming export economy, the Argentine middle classes found that the very success of the traditional export-based economic structure presented serious obstacles to the achievement of their hopes: Argentines could only expect to rise so far in the great foreign-owned concerns, and were left with the alternatives of the limited entrepreneurial opportunities of petty commerce or a career in the government bureaucracy. The "peculiar form of disguised unemployment" that prevailed among this social group, which David Rock appropriately terms the "dependent" middle class, made it fertile political ground for the Radicals. Despite their rhetoric, which promised to vindicate the common man, the Radicals proposed no basic changes in the economic structure but rather envisioned a more equitable distribution of the economy's benefits.[32]

The strategy of the Radicals, under the domineering rule of party leader Hipólito Yrigoyen, was to abstain completely from the political process until the government guaranteed honest elections. Though Yrigoyen seldom appeared in public or delivered speeches, his tireless organizational work and his spartan personal habits fostered the idea that he was devoted to the common people's cause. Portraying himself as the personal embodiment of the movement for effective suffrage and constitutional democracy, Yrigoyen avoided potentially divisive economic and social questions and concentrated the efforts of his party solely on political reform. As a result, the Radical Party attracted a large following during the first decade of the century, particularly among the middle classes, and threatened to support armed revolt if the government failed to reform the electoral process. The party had already made this point clear when it had organized an unsuccessful coup against the regime in 1905.[33]

Confronted with the threat of rebellion from both the anarchosyndicalists and the Radicals, presidents after 1905 began to dismantle the traditional political system. The 1904 election had installed Manuel Quintana, an elite lawyer and landowner, and a friend of General Roca, in the presidential chair. Quintana, however, died the next year, and Vice-President José Figueroa Alcorta, whose relations with Roca and the PAN were less cordial, entered Argentina's Government House, the Casa Rosada. During Figueroa Alcorta's administration the approaching demise of the traditional political system became clear. In the Chamber, Roca's deputies harassed the President and refused to sanction the budget laws for 1908, causing a crisis as the 1907 session neared its end. In response, Figueroa Alcorta issued decrees closing Congress and placing the budget of 1907 into effect for the coming year. Moreover, under a clause in the Argentine Constitution permitting the federal government to assume temporary political control in provinces where irregular electoral situations occurred, Figueroa Alcorta intervened in seven provincial governments and worked to destroy the PAN machine.[34]

With the political system in crisis, Figueroa Alcorta, whose term expired in 1910, searched for a successor capable of enacting electoral reform and integrating the Radical Party into the political system. The man selected to be the candidate of the Unión Nacional, a hastily gathered political conglomerate composed of supporters of Figueroa Alcorta and Pellegrini, was Roque Sáenz Peña, one of the republic's most distinguished statesmen. An experienced diplomat and the son of a former president, Sáenz Peña was serving as ambassador to Italy at the time of his nomination.[35] On his return to Argentina, he immediately made clear his support for electoral reform. After having been elected, and before taking office, he met with Yrigoyen and emphasized his determination to restructure Argentine political life. In turn, Yrigoyen promised that if reforms were carried through, the Radicals would take part in elections.[36] True to his word, the new President gave the highest priority to electoral reform, and in 1912 Congress passed legislation, known since as the "Sáenz Peña Laws," creating a system of fair elections based on the secret ballot and obligatory voting for all male citizens over the age of 18. This was the beginning of Argentina's first period of political democracy, which was to endure until the coup of 1930.

Enmeshed in the complicated political maneuvering that preceded the election of Sáenz Peña, the Figueroa Alcorta government did not give much attention to the formulation of petroleum policy. However, some members of Congress began to urge action. Acting on behalf of farmers who needed cheaper fuel, a group of deputies from the agricultural regions of Santa Fe province introduced legislation in September 1908 designed to enable the government to begin systematic development of its petroleum reserve. The deputies urged the Chamber to authorize the President to spend 200,000 pesos to purchase two new drilling rigs plus storage tanks, pipes, and other equipment. Emphasizing that oil would bring Argentina "an inexhaustible source of wealth and the prospect of colossal progress," they urged the government to act decisively to develop the reserve. A similar discovery in Europe or the United States would have led to rapid development, they claimed; and they pointed with scorn to the government's inaction, which not only postponed further exploration and drilling but also left what oil had been produced to evaporate or seep into the ground.[37]

The fate of this 1908 proposal, which the Chamber of Deputies unceremoniously shelved, points up a second major obstacle that blocked development of the country's oil fields—the influence of liberal economic ideology in Congress, particularly among members representing the landed elite. In 1909, President Figueroa Alcorta decided to give cautious support to the new oil industry and asked Congress to appropriate 500,000 pesos to enable the state to undertake production on a regular basis. He pointed out that studies had confirmed the quality of the oil, which was already being used with great economies on the state's Patagonian railways. To sweeten the pill for defenders of laissez-faire economics, he decreed a drastic reduction in the size of the state reserve to 7,950 hectares around the town of Comodoro Rivadavia; moreover, to placate those opposed to state capitalism, the proposed legislation contained a provision enabling the president to divide the state reserve into 625-hectare plots and to auction them off to the public in the event that the government's operations were unsuccessful after five years.[38]

Figueroa Alcorta's proposal aroused strong antagonism in Congress, particularly in the upper house. Many influential senators were convinced that Argentina owed her prosperity to the liberal international economic system and were worried that government intervention in the oil industry might set an unhealthy precedent and lessen the traditional attractiveness of Argentina to foreign investors. Joaquín V. González, a powerful senator from La Rioja, led the opposition by attacking the president's proposal as an unconstitutional violation of the free-enterprise system. Nonetheless, several senators expressed apprehension lest the state give up an oil reserve whose magnitude was still unknown. In the end the Senate passed the legislation, but not without further reducing the state reserve to 5,000 hectares, a decision ratified by the Chamber of Deputies. Thus Figueroa Alcorta's proposal became Law 7059 of 1910.[39]

Once this legislation was enacted, some 195,000 hectares of the 1907 reserve were opened to private investors, now able to request petroleum concessions from the Bureau of Mines. Investors, in fact, had shown considerable interest in Patagonian oil soon after the original discovery; in 1908, the government received no fewer than 109 requests for exploration permits in Neuquén, Chubut, and Santa Cruz—three of the five national territories that made up Patagonia.[40] This surge of private interest prompted several congressmen to warn that protective legislation was vital to prevent Standard Oil and other "trusts" from acquiring Argentina's best oil lands. But with the passage of Law 7059 in 1910, investors began a speculative land rush in the Comodoro Rivadavia region that obliged the Bureau of Mines to concede title to 42,000 hectares and grant exploration permits over a much larger area by 1913.[41]

The mining code then in effect—dating from 1886—presented no major obstacles to private investors who desired to acquire petroleum lands. Under the prevailing interpretation of the Argentine Constitution, each province enjoyed the right to administer mineral concessions within its boundaries; the national government could administer mineral concessions only in the national territories. The nation or the individual province, under the terms of the 1886 code, was the original owner of all mineral resources (Article 7) and could transfer domain to private parties under procedures outlined in the code (Article 8). Written chiefly with mineral production in mind, the 1886 code contained terms and specified time spans that oil experts considered detrimental to rational oil exploration. The private investor began by applying for a *cateo*, or permit, which gave him the right to explore an area of up to 2,000 hectares (Articles 23–27). But in order to obtain title, the explorer had to begin work within 30 days of receiving the permit and had to complete his exploration and drill at least one well within another 300 days. (A concessionaire who was still drilling when this period ended could request an extension of 15 months, or, in special cases, an extension until oil was discovered [Articles 28–29].) When the explorer verified his discovery, or discoveries, he then could request title over one or more concessions of 600 hectares each—though one individual could not request title over more than three adjoining concessions (Articles 29, 226). If the requested concession was on public land, the government would grant title free; if on private land, the concessionaire would have to arrange purchase (Articles 44–45). Finally, to retain his grant, the owner was obligated to employ at least four workers on it for at least 230 days annually (Article 269).[42]

Because more than 30 days were often needed to install drilling equipment in isolated areas, and because the 300-day exploration period was widely recognized as insufficient for thorough exploration (one to two years being usual), the Bureau of Mines did not enforce the code strictly prior to 1923. In particular, it often disregarded the 300-day limit.[43] Under such conditions of loose enforcement speculation could and did flourish, and groups of individuals associated with a single interest gathered *cateos* covering huge expanses—up to 113,500 hectares in one case—without seriously carrying on exploratory drilling. Speculators of this sort typically delayed exploration by filing lengthy and convoluted requests for exemptions.[44] The mining code, complained José Manuel Méndez, author of one of the first serious treatises on Argentine petroleum in 1916, was "a mixture of restrictions and exemptions that made possible the use of multiple subterfuges to evade the law."[45]

Shortly after Law 7059 took effect, the presidential administration of Roque Sáenz Peña took office; and the new chief executive soon demonstrated his determination to follow a more vigorous petroleum policy than that of Figueroa Alcorta. Though Sáenz Peña generally adhered to liberal economic policies and opposed state intervention in the economy, he believed that oil should be an exception. A number of factors influenced Sáenz Peña's decision to adopt an aggressive oil policy. One was the frequent and prolonged coal strikes that were halting production in Britain during this period; they disrupted the Argentine economy and aroused deep concern in the press. By 1912, the nation's most prestigious newspaper, *La Prensa*, was urging the government to develop its oil industry in order to reduce Argentina's dependence on British coal. Another factor was what one scholar has called Sáenz Peña's reputation as an "outspoken and consistent Yankeephobe." Long a critic of U.S. economic policy in Latin America, the president opposed the expansion of North American "trusts" in the Argentine meat-packing industry and viewed the possible entry of U.S. oil companies into Argentina as a matter of grave concern.[46]

Two months after taking office, on December 24, 1910, Sáenz Peña moved to revitalize the state's fledgling oil operations by reorganizing them into a formal commercial agency. The President issued a decree transferring the administration of the state's oil fields from the Bureau of Mines to a newly created Dirección General de Explotación del Petróleo de Comodoro Rivadavia, hereafter referred to as the Petroleum Bureau. Sáenz Peña's decree established not only the institutional framework for the state oil firm but also its objective, which was to produce and sell petroleum "especially for the Navy and the National Railways." To administer the Petroleum Bureau, which like the Bureau of Mines remained within the Ministry of Agriculture, Sáenz Peña appointed a five-man Administrative Commission headed by Director-General Luis A. Huergo, a distinguished professional engineer. To head operations at the Patagonian fields, Sáenz Peña appointed another mining engineer, Leopoldo Sol, who arrived at Comodoro Rivadavia on April 1, 1911, the day the Petroleum Bureau formally assumed operation of the state oil fields.[47]

The newly appointed administrators found that the Petroleum Bureau faced a multitude of problems: the funds appropriated only the year before under Law 7059 were already largely exhausted; and the four wells in production in April 1911 had had their output reduced by drifting sand. In short, work at Comodoro Rivadavia was nearly at a halt. To revive operations, the commissioners addressed an urgent plea to the government to appropriate two million pesos, the minimum sum they argued was neces-

TABLE 1.2. *Argentine Government Appropriations for the State Oil Industry*
(Paper pesos)

Year	Authorization	Amount
1910	Law 7059, Sept. 6, 1910	500,000
1911	None	—
1912	General budget	1,000,000
1913	General budget	1,500,000
1914	General budget	1,500,000
1915	General budget	1,000,000
	Ministerial resolution, Oct. 14, 1915	1,295,500
	Ministerial resolution, Dec. 30, 1915	859,741
1916	General budget	1,000,000
TOTAL, 1910–16		8,655,241

SOURCE: Frondizi, *Petróleo y política*, p. 68.

sary to drill 20 new wells, clean the old ones, and build an adequate system of supply and storage. The commissioners also pointed out that construction of an aqueduct to supply the oil fields with water from Behr's Estancia was a vital first order of business. It was not solely the quality of the local water supply (though it was certainly terrible, and resulted in a high local mortality rate, especially from typhoid fever) that led the Petroleum Bureau to attach such high priority to the water problem; a steady water supply was essential in order to use modern rotary drilling equipment instead of the primitive cable drills still being employed.[48] Though Sáenz Peña strongly endorsed the Petroleum Bureau's request and argued that the resulting increase in oil production would reduce Argentina's "painful tribute to foreign countries," Congress appropriated only one million pesos for 1912.[49] This sum was barely adequate to permit operations to continue. As Table 1.2 demonstrates, the government's appropriations to support its oil industry totaled only 8.6 million pesos by 1916, when government support ceased and the Petroleum Bureau began to finance itself completely from its own earnings.

Perpetually starved for investment capital by a congressional majority unfriendly to the concept of a state oil industry, the Petroleum Bureau made slow progress in its Patagonian operations. Through 1912 the Comodoro Rivadavia fields remained in a state of suspended animation while the Administrative Commission devoted much of its meager funds to building the essential elements of an operational substructure. The aqueduct received top priority and was completed in May 1913, after Huergo, in a dramatic gesture, pledged to finance it out of his own fortune if necessary. This temporarily resolved the water problem.[50] Of importance, too, was the construction of port and dock facilities at the open roadstead of Comodoro Rivadavia. Most ships had to anchor one or more kilometers off the coast and transfer goods to and from launches, which had difficulty weathering the rough seas in the frequent periods of strong winds. Indeed, when Congressman Nicolás Repetto visited Comodoro Rivadavia in 1914, stevedores had to carry members of his party ashore through the surf—hardly an appropriate arrival for a Socialist politician.[51] To alleviate these costly and time-consuming procedures, the Petroleum Bureau began to construct a pier in 1912, but the project was not completed until the 1920's.[52] In addition to the aqueduct and the pier, the Commission authorized construction of four large storage tanks, some inexpensive workers' housing, and a small refinery to produce heavy crude oil for the navy. The rest of the budget went to pay wages and to repair existing wells and drill a few new ones.[53] Production, which grew slowly in this situation of capital shortage, reached 20,732 cubic meters in 1913.

Distressed at this slow rate of development, Huergo requested early in 1913 a budget of 15 million pesos for the following year to permit exploration of the entire 5,000-hectare reserve, rapid expansion of production, and organization of a tanker fleet. Sáenz Peña supported this budget with a message to Congress emphasizing that prudence, the world political situation, and the need "to create a new industrial state and true economic independence" counseled increasing the state's oil production rapidly.[54] Congress, however, appropriated only 1.5 million pesos—the same amount it had voted for 1913. Later on, in July, Sáenz Peña urged Congress to approve the sale of 15 million pesos of special oil-development bonds, but the legislators pigeonholed the project.[55]

In the meantime, Huergo attacked the aimlessness of national oil policy in a vitriolic manifesto designed to marshal public support behind the Petroleum Bureau by raising the specter of Standard Oil as a monopolistic "trust" out to seize control of Argentine oil. This manifesto marked the opening salvo in the long and bitter battle the Petroleum Bureau and its successor waged against the international oil companies, particularly Jersey Standard. Luis A. Huergo enjoyed great prestige in Argentina. A pioneer in Argentina's engineering profession, Huergo had led a struggle during

the 1880's to build a low-cost port for Buenos Aires with national capital
and Argentine engineers. Instead, the government had chosen to build a
port designed and financed by the British; that it proved cumbersome and
very expensive both to construct and to operate only heightened Huergo's
already distinguished professional reputation. When Sáenz Peña appointed
him to head the Petroleum Bureau, he was at the pinnacle of his career.[56]

Huergo directed the bulk of his 1913 statement, which he first sent to
the Minister of Agriculture and then made public in April, against the
concession policy of the Bureau of Mines. Pointing out that private investors, who Huergo believed were mainly linked to foreign interests, had
acquired thousands of hectares of oil land around the state reserve, the
Director-General warned that foreign powers were on the verge of seizing
control of the republic's oil resources. Huergo criticized the government's
oil concession policy as openly contradicting the intent of the mining code,
noting that some private investors had held exploration permits for as long
as two years without performing exploratory work. Argentina should avoid
the fate of Mexico—where foreign oil firms controlled the industry and,
Huergo claimed, intervened in the social revolution then under way by
financing leaders friendly to their interests—by revising its petroleum
policy. Though Standard Oil did not begin to acquire Argentine oil lands
openly until 1920, Huergo was convinced that its agents already were attempting to monopolize Argentina's best reserves. He attacked the giant
U.S. corporation vigorously: "Throughout the world," he wrote in a passage often quoted in Argentina, "Standard Oil acts like a band of cruel,
usurious pirates, headed by an ex-clerk, who began by carrying thousands
of families among his own countrymen to ruin. Like an octopus, [Standard
Oil] has extended its tentacles everywhere, accumulating colossal fortunes
of millions of pesos on the basis of human blood and tears."[57]

Huergo died on November 4, 1913, some six months after publishing
his manifesto, but it had immediately attracted public attention and induced the Sáenz Peña government to strengthen its oil policy. The Chamber of Deputies had quickly appointed a special committee to investigate
Huergo's charges of lax enforcement of the mining code; but when it ended
by doing little, Sáenz Peña had utilized his executive powers under the
1903 Public Lands Law to reform concession policy. He had issued a decree
on May 9, 1913, requiring strict compliance with the mining code, declaring null and void all concessions not being worked, and greatly enlarging

the state's reserve by about 160,000 hectares in Chubut territory. This decisive action firmly established Sáenz Peña's reputation as a precursor of Argentine petroleum nationalism.[58]

Political circumstances suddenly became less favorable for the state oil industry when Sáenz Peña left the presidency for health reasons late in 1913. His successor, Vice-President Victorino de la Plaza, was a staunch economic liberal and opponent of state capitalism who moved within a year to modify Sáenz Peña's reserve policy. By a decree of October 8, 1914, de la Plaza limited the applicability of the 1913 decree to publicly owned lands within the enlarged reserve. Private investors now were able to acquire major new petroleum concessions, and concessions in fact increased from 16,393 hectares in 1913 to 99,619 in 1914. De la Plaza also asked Congress for permission to turn over the operation and administration of the state's reserve to private interests.[59]

Yet despite the poor record of Congress in supporting the Petroleum Bureau, de la Plaza's proposal stirred controversy in the Chamber of Deputies. Indeed, a similar proposal introduced by Conservative deputies had died there in 1913. One group of deputies, led by Radical Alfredo Demarchi, president of the Unión Industrial Argentina, the republic's principal association of industrialists, argued that an alternative existed to private ownership on the one hand and complete state control on the other. Demarchi condemned de la Plaza's proposals as tantamount to delivering the oil fields into the hands of a foreign "trust," and urged Argentina to form an enterprise based on the mixed-company model the British government had successfully developed when it formed the Anglo-Persian Oil Company. The shares of such a company would be held both by the government and by private (presumably Argentine) investors. Foreshadowing a line of argument Argentina's small but growing industrial class would long hold to, Demarchi maintained that unrestricted economic liberalism would enable foreign companies to establish nonproducing oil reserves in Argentina, whereas a mixed-company arrangement would assure efficient production of oil for the republic's "industrial future." Because of the diversity of views regarding oil policy and the fragmentation of the party system in Congress between 1912 and 1916, the Chamber of Deputies took no action on any oil-legislation proposals made during the de la Plaza presidency. Congress continued state operations at Comodoro Rivadavia at low levels of funding and postponed policy decisions until after the 1916 national elections.[60]

While the Congress in Buenos Aires delayed action and the clouds of approaching political and economic crisis gathered over Europe, at Comodoro Rivadavia the Petroleum Bureau struggled to increase production in the state oil fields. Yet antiquated equipment, shortages of drilling machinery, and a scarcity of skilled labor continued to frustrate the bureau's efforts. By mid-1914, 13 wells had been completed at Comodoro Rivadavia, but only seven produced oil. Owing to its woefully inadequate finances, management complained that "practically no new exploration has taken place."[61] A shortage of ocean transport meant that at first only sporadic shipments reached Buenos Aires; the oil held at Comodoro Rivadavia still was stored in open-air cisterns. To correct this wasteful situation, in 1914 the Petroleum Bureau leased one small tanker and borrowed another from the navy to establish regular shipping patterns.[62] Partly because the state set a low price on it, Comodoro Rivadavia petroleum made a favorable impression when it did appear in the Buenos Aires market. Regular sales of crude began with the first of the new tanker shipments, and demand immediately outstripped supply. One important customer was the Compañía Italo Argentina de Electricidad, a large utility that had experimented with state oil as early as 1912, and other buyers included breweries and flour mills. The largest customer of all, the Ministry of Marine, purchased over 20,000 cubic meters of oil in 1914.[63]

Thus Argentina had attained a unique position among oil-producing countries by the time the First World War began, for she boasted the world's first state-owned oil company. As we have seen, however, bitter political conflict had surrounded the company from the start, and the debates about the proper role of the government in the economy were far from over. Yet during the war Argentina would undergo momentous changes that would strengthen support for the government petroleum enterprise, as we shall see in the next chapter.

CHAPTER II

Energy Crisis: Petroleum Policy During the First World War,
1914-1919

When Hipólito Yrigoyen rose to power in 1916 as Argentina's first demo-cratically elected president, he faced the worst economic depression of modern Argentine history. Though the republic remained neutral in the First World War, the European conflict crippled the Argentine economy and produced an energy crisis of severe proportions. In this context, rapid development of the republic's petroleum resources, particularly the state oil fields at Comodoro Rivadavia, became a serious public issue. But the Yrigoyen government failed to respond decisively to the challenge of the wartime fuel crisis, and the state petroleum enterprise grew at a disappoint-ing rate. This wartime emergency was of the utmost significance in the evolution of Argentine petroleum nationalism, for the republic's dismal economic situation and the resulting social unrest prompted both military strategists and an influential group of intellectuals to analyze Argentina's dependent economic system critically and to conclude that a more self-sufficient, industrialized economy was essential. During the 1920's these men, particularly Colonel (later General) Enrique Mosconi, would build an ideology of state petroleum development based on this criticism and would support their argument that the nation's future prosperity and se-curity required a flourishing, government-owned oil industry by pointing to the economic crisis caused by the war.

The Economic Impact of the War

The end of Argentina's era of heady prosperity began in 1913, when the Balkan War crisis sharply reduced new European capital investment. The decline of capital flow, along with the failure of the 1913–14 crop, led to serious balance of payments deficits. Gold rapidly flowed out of the country, and in response, the supply of paper money in circulation, tied by law to the amount of specie in the banks, fell rapidly. A forced liquidation took place, and by the time the First World War broke out the Argentine economy was virtually in a state of panic. After July 1914, shipping became scarce, foreign trade was paralyzed for weeks, and commercial life dwindled. As exports plummeted, imports also fell (by 30 percent in 1914), thus reducing the government's income, which flowed largely from customs duties. To halt the gold outflow and the resulting currency deflation, on August 2, 1914, the de la Plaza government took Argentina off the gold standard by a decree that ended conversion of paper money into specie.[1]

During 1914 and 1915, severe depression hit major sectors of the Argentine economy, particularly the construction industry, which suffered a near-total collapse (see Table 2.1). Import prices doubled on the average between 1910 and 1917. Touched off in part by the soaring prices of imported goods, on which such a large proportion of mass consumption depended, inflation began and plagued Argentina throughout the war. As a result, the real income of the working classes fell, while at the same time unemployment mounted—to reach nearly 20 percent of the work force by 1917.[2] Misery struck much of the Argentine population. Long lines formed at soup kitchens, and severe shortages of basic consumption items such as cooking oil occurred. In the heart of Buenos Aires, near the Retiro railway station, a huge shantytown sprawled; in the suburbs, people scoured garbage dumps for paper, bottles, tin cans, or anything else they might be able to sell. Unemployment and inflation struck cities and towns throughout the interior as well. In Catamarca, wages rose only 5 percent in the period 1914–24, but prices more than doubled. Everywhere the soup kitchen became a symbol of the economic impact of the war.[3]

Gradually, the export trade revived, stimulated by European war demands. Nonetheless, it was not until 1918 that the value of exports rose significantly above the 1913 level. Lagging export volume reflected a series

TABLE 2.1. *The Impact of the First World War on the Argentine Economy: Indexes of Leading Economic Indicators, 1913–19*

Economic indicator	1913	1914	1915	1916	1917	1918	1919
Exports	100	77.6	112.1	110.4	106.0	154.4	198.6
Imports	100	65.0	61.6	73.8	76.6	100.9	132.1
Building permits[a]	100	40.1	20.9	16.0	15.6	23.3	31.2
Industrial output	100	91.0	81.6	83.9	83.0	99.1	103.1
Consumer prices[a]	n.a.	100	107.0	114.8	135.0	169.4	159.6
Real wages[a]	n.a.	100	93.4	80.3	68.8	62.0	83
Unemployment (percent)[b]	6.7%	13.7%	14.5%	17.7%	19.4%	10.3%	7.9%

SOURCES: Export and import figures are from Tornquist, *El desarrollo económico de la República Argentina*, p. 134; and Williams, p. 54. Figures for building permits are from Tulchin, "The Argentine Economy During the First World War," p. 44. Industrial output figures are from Di Tella and Zymelman, p. 309. Consumer price figures are from Alejandro Bunge, *The Cost of Living in the Argentine Republic*, p. 7. Figures for real wages are from Di Tella and Zymelman, p. 317; and Alejandro Bunge, *The Cost of Living in the Argentine Republic*, p. 7. Unemployment percentages are from Tornquist, *El desarrollo económico de la República Argentina*, p. 15.

[a] In Buenos Aires.

[b] The figure for unemployment is the percentage of the labor force in Buenos Aires unemployed in August of each year, except for 1918, when the figure is for March.

of tragic crop failures; in fact, Argentina did not have a single bumper crop during the entire war. Severe shortages of shipping further disrupted the export trade, and the total tonnage of foreign shipping entering and leaving Argentine ports dropped by almost half between 1913 and 1918. In 1917 and 1918, much of the corn crop rotted for lack of shipping space. Beef, exports of which more than doubled between 1914 and 1918, represented the single bright spot in Argentina's foreign trade.[4]

Following the collapse of 1914 and 1915, imports revived more slowly than exports, and as late as 1917 the value of imports was only about three-quarters what it had been in 1913. Great Britain, Argentina's traditional source of manufactured products, was unable to supply more than a trickle of goods during the war, and imports from the United States also became scarce. Isolated from traditional sources of supply, the Argentine economy suffered severe shortages of fabricated metal products, railway equipment, and machinery of all sorts.[5]

Though it is often assumed that the Argentine manufacturing industry boomed during the war in response to the import shortage, in fact, as Table 2.1 shows, industrial output fell during the early years of the war and did not recover to prewar levels until 1918. H. O. Chalkley, British commercial attaché at Buenos Aires, effectively summarized the plight of the manufacturing sector when he reported that "nearly all Argentine industries have been adversely affected by scarcities." Many small industrial firms, particularly hard hit by shortages of imported capital goods and raw materials, were forced to cease operations. Some industries that utilized Argentine-produced raw materials expanded (for example shoe manufacturing, woolen textile manufacturing, and the chemical industry), but even in these cases the lack of skilled workers and the virtual impossibility of importing machinery impeded really rapid growth.[6]

The fuel shortage dealt staggering blows not only to industry but to the entire economy (see Table 2.2). Coal imports fell disastrously, hampered by shipping shortages and European military demands, and the price of coal rose at an alarming rate. In 1918 coal sold in Buenos Aires at an average price of 50 gold pesos per ton, an increase of over 500 percent since 1913.[7] As the coal situation grew bleaker, power plants, meat packers, and other major consumers tried to shift to oil; but petroleum imports were scarce, national production grew slowly, and oil prices consequently rose steeply after 1915.[8] By 1918, the fuel shortage had created a national economic

TABLE 2.2. *The Impact of the First World War on Argentina's Fuel Supply*

Fuel	1913	1914	1915	1916	1917	1918	1919
Coal							
Imports (metric tons)	4,046,278	3,421,526	2,543,887	1,884,781	707,712	821,974	1,000,000
Price per metric ton (gold pesos)	9.3	n.a.	n.a.	14.0	30.0	50.0	40.0
Oil							
Imports (m³)	280,000	235,000	460,000	365,000	355,000	185,000	540,000
Domestic production (m³)	20,733	43,795	81,580	137,551	192,374	214,867	211,300
Price per m³ (paper pesos)[a]	n.a.	23.3	21.8	41.8	62.0	82.9	62.4
Wood							
Domestic production (metric tons)[b]	1,205,565	2,451,465	1,865,739	2,764,485	2,853,337	n.a.	n.a.
Price per metric ton (paper pesos)	26.0	25.0	23.5	27.0	31.3	42.5	n.a.

SOURCES: Figures for coal imports and prices are from Tornquist, *El desarrollo económico de la República Argentina*, p. 92; and Havens, p. 649. Figures for oil production are from Dorfman, p. 143; for oil imports, from "Resumen estadístico de la economía argentina." *Revista de Economía Argentina*, 20 (Nov. 1938), p. 326; and for oil prices, from Argentina, [4], vol. 1, facing p. 348. Figures for wood production and prices are from Tornquist, *El desarrollo . . .*, pp. 93–94.

[a] These figures represent weighted indexes of prices fixed by the private companies and by the state oil firm.

[b] These statistics are for tons of wood transported on the railways; actual production was somewhat higher, since some wood was transported by water.

emergency: some factories were forced to shut down, electric power plants had to reduce output and increase prices to triple 1913 levels, and the railways were increasingly unable to maintain normal operations. By late 1917, the London directors of the Argentine railway companies were warning their government that the coal shortage would soon force Argentine food exports to drop.[9]

The railways reluctantly turned to burning wood. But wood prices also rose, and the *quebracho* and *algarrobo* from the northern parts of Santa Fe and Santiago del Estero provinces provided energy far less efficiently than coal.[10] Nonetheless, by 1916 wood already supplied about 70 percent of the railways' fuel; and by 1918, wood accounted for almost 75 percent of the republic's total fuel supply.[11] By then, however, wood was also becoming scarce, and both power plants and the railways had to resort to burning corn. Katherine Dreier, a perceptive U.S. visitor to Argentina, related that on a 1918 rail journey from Mendoza to Buenos Aires she "was much interested to see that ears of corn were fed into the engine." The train's engineer informed her that corn fuel was "not good for the engine. . . . Extracts of oil . . . made it necessary for the engine to be constantly overhauled." In the republic's power houses, according to fuel expert V. L. Havens, a great deal of corn was consumed during the war, despite its low heat-producing value.[12]

The energy crisis produced serious rumblings of discontent in the Argentine military during the First World War. Though the army did not suffer directly, by the end of the war a number of army officers were arguing that dependence on imported fuel was strategically dangerous and a threat to national security. The navy was far more seriously affected by the fuel crisis than the army, for its ships still primarily burned coal. Only two of its 15 capital ships, the recently completed battleships *Rivadavia* and *Moreno*, were equipped to burn oil. And though 12 of the fleet's 15 destroyers could use oil, all of the smaller ships and support craft relied exclusively on coal.[13] Cut off from traditional suppliers, the Ministry of Marine was obliged to reduce its coal consumption sharply in order to retain sufficient stockpiles for the future. Despite a reduction in consumption from 1,100,000 metric tons in 1911–15 to 600,000 in 1916–20, the navy's fuel shortage had reached crisis proportions as early as 1917.[14] To keep the fleet in operation, the government in that year approached industrialists in an attempt to exchange state oil for stockpiled coal. Only the Armour meat-packing plants responded, and the navy's distress deepened.[15]

Appalled by the traditional official disinterest in exploiting national coal resources, the Ministry of Marine decided in 1917 to mount its own exploration campaign "with the purpose of finding deposits of importance for the fleet." The navy soon located good coalfields in the Andean foothill province of San Juan and in Patagonia, in Santa Cruz territory. Tests of the coal's quality in a navy laboratory showed that Argentine coal was of at least as high quality as Chilean coal, which had long been used by the Chilean fleet. Ministry officials and naval officers attacked earlier "official publications" that had downgraded the quality of Argentine coal, and stressed the immediate need for a mining development program.[16] The government, however, insisted that mineral exploration was the province of the Bureau of Mines and the Ministry of Agriculture, and refused to permit the Ministry of Marine to undertake coal mining. Enrique Hermitte, director of the Bureau of Mines, confirmed that good coal samples had been found in Patagonia but complained that the government had appropriated no funds to promote exploration and development. "It is with real distress," Hermitte wrote, "that I see the field instruments laid up on their shelves and the technical staff sitting at their desks, owing to lack of funds to send them out."[17] Because of the government's inertia, Argentine coal production remained insignificant and reached a wartime peak of about 5,000 metric tons in 1918.[18]

Frustrated in its quest to promote coal exploration, the Ministry of Marine turned its attention to oil. Since 1914, the Petroleum Bureau had been operating a small refinery at Comodoro Rivadavia to prepare fuel for the fleet. Convinced of the utility of oil fuel, the navy began by 1916 to promote rapid development of the state oil fields. In Buenos Aires, navy officers gave public lectures on the relationship between oil development and national defense, and the Ministry of Marine lobbied busily to convince the government to increase petroleum production. "The fuel problem is perhaps the most difficult question our country faces," emphasized Federico Alvarez de Toledo, Yrigoyen's civilian Minister of Marine, in 1918. "In order to navigate our ships and operate our power plants, the development of our *very rich oil deposits* must not be delayed."[19] Indeed, Alvarez de Toledo went so far as to suggest that "naval functionaries" should control the governments of Patagonia to construct the modern ports the region required, to administer the large foreign-born population, and, most importantly, to stimulate rapid development of oil production.[20]

The Rise of Economic Nationalism

The wartime crisis that reverberated so strongly in the navy also sparked a critical examination of the nation's economic structure among a small but influential group of Argentine intellectuals. The essayists and economic theorists who challenged the soundness of the traditional export-oriented economy made a profound impact on the evolution of twentieth-century Argentine economic thought. Their fascination with industrial power and economic independence places them among the intellectual precursors of modern Argentine economic nationalism.

"The war has demonstrated . . . our lack of real independence," lamented one of these writers, Enrique Ruiz Guiñazú, in 1917. His comment aptly summarized the basic premise of these early economic nationalists, that prewar Argentina had been excessively dependent on foreign capital, markets, and imports. To reduce this dependence, the government was urged to enact a protective tariff, to finance an infrastructure that would stimulate economic diversification and development of the interior's resources, and to provide the private sector with credit and vocational-education facilities. Unlike many economic nationalists after the Second World War, these early critics of the liberal international economy did not call for extensive government ownership and state control; their main goal was to promote industrialization and national economic self-sufficiency within the framework of private capitalism.[21]

Japan's spectacular industrial growth was one model that impressed the Argentine critics of dependence who appeared during the First World War. The brilliant young essayist Manuel Ugarte, for example, whose outspoken attacks on U.S. influence in Latin America brought him fame during the 1920's, argued that excessive reliance on foreign trade and investment were "factors of denationalization" and economic weakness in Argentina, and urged his countrymen to build their economic system along the lines of "that marvelous Japanese empire," which had "learned the secrets of all traditions and combined them into its own system, without alienating any of its own future."[22]

Imperial Germany provided perhaps the most influential model of nationalist development to this generation of Argentine economists. In a provocative essay, the contemporary American economist Harry G. Johnson has traced the proindustrial and protectionist policies that have taken strong root in many underdeveloped countries to the "nationalist and interventionist" ideas the nineteenth-century German economist Friedrich List set forth in his *National System of Political Economy*. List's theories, which emphasized that protection of a native manufacturing industry against foreign competition would provide a market for agriculture and promote sustained national economic development, strongly influenced twentieth-century Argentine economic thought. Alejandro E. Bunge, Argentina's foremost theorist of protectionist economic development, vigorously promoted List's ideas as the basis for a viable model for national development during and after the First World War.[23]

Bunge, who was educated in Germany and became Argentina's leading economist by the mid-1920's, constructed a powerful critique of the republic's economic dependence.[24] He was perhaps the first Latin American economist to analyze economic underdevelopment systematically in terms of excessive structural reliance on the developed countries. The First World War, he wrote in 1917, demonstrated that the export economy invited disaster by binding the republic's economic life too closely to foreign markets and capital. Argentina, Bunge lamented, had become a "satellite" of the industrialized countries. A fervent advocate of a protective tariff, of government programs to stimulate development of the interior, and of the expansion of mining and oil production, Bunge enjoyed great prestige during his long career as university professor and government adviser. His ideas strongly influenced Argentine and Latin American economic thought toward industrialization and self-sufficiency.[25]

Bunge created a framework for the analysis and criticism of the export-dependent economy, but his major policy recommendation—a protective tariff to spur industrial growth—did not attract strong political support before the Great Depression except from industrialists, who remained politically weak, and from some provincial political leaders, military officers, and intellectuals.[26] Despite the wartime crisis, the traditional liberal economic system still enjoyed broad support among the urban population, long accustomed to European imports and opposed to the tariff as an unnecessary tax on consumers. In this sense, urban consumers shared with export-oriented landowners a common commitment to Argentina's historic policy of free trade. This consensus endured until the Great Depression shattered the international trading system. During the economic crisis of the 1930's and the Second World War, what some analysts call "popular" or "left-wing" economic nationalism gained strong political support, par-

2. Alejandro E. Bunge in his office, June 1930.

ticularly among the urban middle class. Though this ideology set forth goals similar to those of Bunge, it advocated more extreme measures to achieve them. A sweeping ideology of national development, it aimed to promote national economic power and the welfare of the masses by means of extensive state capitalism, economic planning, and nationalization of foreign investments; it would become a major force in Argentine politics, particularly during the presidency of Juan Perón.[27]

Petroleum nationalism, one of the major themes of modern Argentine economic nationalism, had its own distinct origins in the writings of the vigorous and popular journalist Ricardo Oneto during the First World War. His numerous essays, which appeared in such newspapers as *La Mañana*, *El Radical*, and *La Epoca*, emphasized that government ownership of oil was essential to protect Argentina's petroleum resources for the nation's industrial future. Again and again, Oneto warned his readers that Argentina's economic welfare and political independence were in peril unless the government adopted petroleum nationalism as official policy.[28]

The government's failure to adopt a successful oil-development policy formed the backdrop for the emergence of petroleum nationalism during the First World War. When the war broke out, President de la Plaza was openly hostile to the whole concept of a state oil company. As fuel shortages became more severe in 1915, though, de la Plaza shifted his position but found that he could not move the Conservative-dominated Congress. Early in 1916, de la Plaza sent a message to Congress portraying the plight of the Comodoro Rivadavia fields: of the state's 5,000-hectare reserve, only 3,300 hectares had been explored, even partially; and because of the severe shortage of equipment, the Petroleum Bureau had been able to drill only 46 wells, 19 of which actually produced oil. Given this sorry state of affairs, the president asked Congress to approve an internal loan of 16 million pesos to enable the state oil industry to expand production to 400,000 cubic meters annually by 1918. He also proposed a national petroleum code to regulate land speculation. But the Chamber shelved both of these measures, and Argentine petroleum policy continued to drift.[29]

Yrigoyen's First Government: Political and Economic Policies

In March 1916, Argentine voters went to the polls in the first nationwide elections held under the terms of the Sáenz Peña reforms. Attention focused on the closely contested presidential race, from which Radical Party leader Hipólito Yrigoyen emerged victorious. For the next 14 years, until his fall from office in the 1930 coup, Yrigoyen dominated Argentine politics more completely than any of the country's twentieth-century leaders except Perón. After 1916, Yrigoyen continued to exercise the same tight control over his party that he had during the previous 20 years. Moreover, by blending timely concessions to the working classes with populist and nationalist appeals, Yrigoyen retained the near-mystical reputation of a popular champion he long had cultivated. To understand the development of the state oil industry between 1916 and 1930, it is essential to analyze his political and economic policies.

When Yrigoyen assumed the presidency, his political position was weak and his party controlled neither house of Congress. Because his political opponents, primarily the Conservatives and the Socialists, were thus able to shelve several of the social and economic reforms he proposed during the opening years of his term, Yrigoyen dedicated much of his first presidency to obtaining congressional majorities. The Radical Party, which held only 44 of the 116 seats in the Chamber of Deputies in 1916, did not gain a majority in the lower house until 1920 and never enjoyed a majority in the

Senate before 1930.[30] Preoccupied with building his party's strength, Yrigoyen devoted much of his first presidency to political questions. He employed patronage liberally to strengthen the party among the urban middle classes, and intervened in the provinces whenever possible to turn out opposition-controlled governments accused of electoral irregularities and to hold new elections more favorable to Radical candidates. Casting the Radicals as the forces of righteousness in a battle against an allegedly evil "oligarchic regime," Yrigoyen carried out 20 provincial interventions during his six-year term, more than any previous Argentine President had.[31] Congress sanctioned only five of these interventions; the rest Yrigoyen imposed by decree when the legislators were not in session. By 1922, the interventions had helped the Radical Party assume control of all the provincial governments except those of Córdoba and Corrientes.[32]

The President's provincial intervention policy, together with his strongly personalistic style of government, alarmed the opposition, which branded Yrigoyen a *caudillo*, the term used in the nineteenth century for an authoritarian leader with a popular power base. Like the *caudillos*, Yrigoyen was convinced that he possessed special insight into governing and that he alone was equipped to decide policy and to find the right answers to every question. Influenced by the nineteenth-century German idealist philosopher Karl Christian Friedrich Krause, a writer much in vogue in Spain and Spanish America during Yrigoyen's youth, the president invoked *krausismo* to justify his personalist rule. He was, he claimed, "predestined by God" to lead a "political apostolate" in which he would govern paternalistically in the interests of all classes and groups.[33] But in the years before 1920 when the opposition parties controlled the Chamber of Deputies, Congress was not impressed with the President's claims to a divine mission and emasculated most of his proposed legislation. More than a few legislators grew suspicious of Yrigoyen's intentions when he broke tradition by failing to appear in person to deliver the annual message opening Congress, by refusing to allow his ministers to submit to congressional interpellations of administration policy, and by treating prominent members of opposition parties like political pariahs.[34]

The domineering style Yrigoyen displayed in his provincial intervention policy also characterized his administration of the national government. He insisted on deciding all important matters personally, which often delayed transaction of government business. For the most part, he chose his ministers on the basis not of professional competence but of political allegiance, allowing them little latitude and governing, as *La Prensa* complained, practically without them. "They are really not Cabinet Ministers but private Secretaries, having little if any opinion of their own, merely doing his [Yrigoyen's] bidding," reported U.S. Consul White in an assessment that many Argentines would have shared.[35] By carefully dispensing favors and patronage, and by employing espionage to ensure the loyalty of government employees, Yrigoyen turned the administration into his personal political machine.[36] But his obsession with politics and administration absorbed so much of his time that he turned his attention to economic and social questions only when crises arose.[37]

Though historians sympathetic to the Radical Party often defend Yrigoyen's first administration as a nationalist-oriented government that maintained neutrality in the First World War despite intense Allied pressure, much of Yrigoyen's nationalism was purely rhetorical. As Waldo Frank once put it, Yrigoyen "shrewdly fed the visionary legend of Argentina as a solitary, prophetic nation, above the battle."[38] Whereas the President's wartime policy statements emphasized the sovereign equality of nations and Argentina's determination to conduct an independent foreign policy, his economic policies were emphatically pro-British and closely followed the tradition of liberal internationalism. To keep the "British connection" intact, Radical economic policy aimed (1) to keep the British market open to Argentine rural exports, particularly meat; (2) to maintain a low-tariff trade policy that would permit a strong flow of imports from Great Britain; and (3) to continue to welcome British investment. This confirmation of Argentina's traditional economic policy allied Yrigoyen squarely with exporting interests—particularly cattle interests, the one major sector of the Argentine economy that thrived during the war.

Cattle-raisers enjoyed strong representation within the Radical government. Although Yrigoyen's background was humble, he had acquired a number of ranches in Buenos Aires province during his political career.[39] This was typical of the party's leadership, the bulk of which, according to sociologist Gilbert Merkx, recently had risen into the upper economic class and acquired lands and herds. Of course a number of traditional landed aristocratic family names, including Alvear, Herrera Vegas, and Pueyrredón, also appeared among the Radical leadership.[40] But whether *nouveau riche* or aristocratic, many of the party's prominent figures were closely associated with the business of raising cattle. Yrigoyen himself was a mem-

3. President Hipólito Yrigoyen (front, center) with Dr. Joaquín C. de Anchorena, President of the Sociedad Rural, leaving the National Cattle Show Auction, 1920. On the extreme left is Minister of Agriculture Demarchi; 2d from left, just behind Yrigoyen, is José Luis Cantilo.

ber of the Sociedad Rural, the exclusive cattlemen's association that was also one of the republic's most powerful interest groups, and five of the eight members of his first cabinet also were members. In Congress, 70 percent of the initiatives made in favor of cattle interests between 1916 and 1922 originated with Radical Party members.[41]

On the other hand, Yrigoyen alienated the republic's farmers by warmly backing wartime trade agreements highly favorable to the British. The first such agreement, proposed by Great Britain and France and approved by Argentina early in 1918, granted the two Allies a credit of 200 million gold pesos for the purchase of wheat, corn, linseed, and oats at predetermined prices that was to be paid off over a period of several years. Congress approved this agreement with little opposition, for the Allied powers, who controlled ocean transport, were after all the only possible buyers. Nonetheless, the agreement, as *The Review of the River Plate* pointed out, "constitutes a breach of neutrality and makes this country an adherent (and a very interested one) of the Allied cause."[42] Argentine farmers, most of whom were unnaturalized immigrants and could not vote, condemned the trade agreement as setting minimum prices far below world market prices and as primarily benefiting grain merchants and exporters.[43]

Firmly committed to maintaining the "British connection," Yrigoyen made no attempts to forge new directions for Argentine economic policy. He showed little interest in the reform and development of the republic's inefficient agricultural sector, and he was hardly more favorably disposed toward industrialists. During the economic crisis following the war, when industrial interests clamored for tariff protection to enable them to compete with reviving foreign imports, Yrigoyen moved with the utmost caution and showed himself in favor of protection only for traditional industries that processed rural products.[44] The government's attitude toward industrialization is revealed in U.S. Consul White's report of a conversation with Honorio Pueyrredón, Minister of Agriculture (and thus in charge of the state oil fields) until 1918:

Doctor Pueyrredón said he thought it would be a great calamity for Argentina to have a very much larger population or to become an industrial country, that the future of Argentina lies in the exportation of meat and still more meat, and that the larger the population, the more will be consumed within the country and the less will be available for export. . . . [If] Argentina became an industrial nation, he did not think she would be able to export manufactured goods in competition with other nations; . . . even if she should be able to a certain extent to supply her own needs, the exchange of commodities would be effected totally within the country and so would be of no benefit. He was particularly unwilling to see that new riches could be created in the country by the development of its own resources.[45]

This official hostility to industry and faith in revived export prosperity illuminate the government's failure to promote full-scale development of its oil fields at a time of national energy crisis. The bleak wartime economy, which severely limited Argentina's financial, industrial, and technical resources, certainly restricted the government's ability to act; but even taking this circumstance into account, it is clear that Yrigoyen's oil policy during his first term lacked strong commitment to the idea of state petroleum development.

Characteristically, the President refused to make any decisions on the future of the government oil fields until he could personally inspect the operations at Comodoro Rivadavia. A visit planned for 1917 was interrupted by a cabinet crisis, and Yrigoyen did not make the trip to Patagonia until the following year. In the meantime, the President confined his oil

TABLE 2.3. *Financial Results of the State's Oil Operations, 1911–20*
(Paper pesos)

Year	Sales	Expenses	Reserves and amortization	Profits	Profits as percent of sales
1911–15	2,182,214	1,810,284	169,526	202,404	9.3%
1916	5,452,495	2,667,217	513,139	2,272,139	41.7
1917	10,048,361	2,564,127	2,484,234	5,000,000	49.8
1918	14,888,867	2,996,661	1,892,206	10,000,000	67.2
1919	11,715,459	3,441,808	2,273,651	6,000,000	51.2
1920	12,803,821	3,840,278	2,963,543	6,000,000	46.9

SOURCE: Argentina, [15], p. 21.

policy to financial reorganization, ending the policy established by Figueroa Alcorta of supporting the Petroleum Bureau through funds from the government's budget. Beginning in 1916, the President proposed no additional budgetary support for the state oil fields and notified the Petroleum Bureau that it would have to rely on its own profits to generate funds for further development. Profits were indeed rising rapidly—they reached five million pesos in 1917, and ten million in 1918 (see Table 2.3)—but they still were inadequate to finance the rapid production increases needed to ease the fuel crisis.[46]

Yrigoyen briefly supported the idea of floating an internal loan to raise funds for development projects, one of which would have been the oil industry. In December 1916, Treasury Minister Domingo E. Salaberry introduced legislation to authorize a 100-million-peso loan to be used to establish a National Agrarian Bank and a merchant marine, and to develop the state oil fields. The 16 million pesos projected for oil development would have allowed the sinking of 40 new wells as well as the purchase of two new tankers, storage facilities, and field transport equipment.[47] The Chamber of Deputies approved the proposal in mid-February 1917, but before the Senate could act the administration switched course and abandoned support of the proposed loan, claiming that the high cost of borrowing money had made it impractical. After that, Yrigoyen made no further move to provide financing for the hard-pressed oil fields.[48]

Turmoil at Comodoro Rivadavia

Underfinanced and lacking support from the administration, the state oil enterprise made slow progress. Production rose about 45 percent between 1916 and 1919, when it reached 188,112 cubic meters. This was unimpressive, considering the rapid strides in oil production other countries were making during these years. One scholar lamented in 1920 that at the current rate of increase it would be another quarter century before Argentina produced enough oil to substitute for even a third of her coal consumption.[49] The government's refusal to finance expansion of the state oil fields except through the Petroleum Bureau's profits had another unfortunate result—labor unrest. Anxious to reinvest as much as possible of its profits, the Bureau kept wages low and neglected the living and working conditions of its laborers. This policy spawned frequent and sometimes lengthy strikes that kept the oil fields in an uproar for years and resulted in large reductions in output.

The state's Comodoro Rivadavia fields were becoming a significant employer. The labor force, which had included only a few dozen men when the Petroleum Bureau began operations in 1911, increased to a total of 1,719 workers and supervisors by the end of 1919. In 1917, only 3.3 percent of the 1,401 workers were Argentine citizens; the rest were Europeans, easily recruited from among the masses of unemployed immigrants in Buenos Aires (see Table 2.4). Yet of the 161 supervisory employees in 1917, 48 percent were Argentine nationals.[50] Many of the foreign workers spoke no Spanish, and Comodoro Rivadavia presented a highly cosmopolitan scene. As late as 1925, according to one observer, "drillers from the Carpathians and the distant steppes of Russia mingle and hobnob with German and Polish drillers from the Baku and the Galician fields."[51]

Whatever their national origins, most workers found Comodoro Rivadavia "a hellish place" and one "to try men's souls," as an American visitor described it. The hostile environment made life in the isolated Patagonian oil town dreary and often dangerous.[52] The fierce winds that blew constantly made men's ears buzz until their nerves cracked, and turned many workers into "broken-down old men" who fled Patagonia for easier climes. Frequent, thick sandstorms stalled trucks and sometimes trapped drivers, as H. W. Wilson, a British diplomat who toured the region in 1922, testified when he arrived sore-faced and exhausted at Comodoro Rivadavia after a harrowing, windswept automobile voyage from Río Gallegos. In this bleak environment many workers turned to using opium, hashish, and heroin, and the contraband drug trade prospered at Comodoro Rivadavia.[53]

As far as wages were concerned, the Buenos Aires press reported in 1917 that the great majority of unskilled oil laborers earned between 2.75 and 3.50 pesos for a working day that averaged 12 hours (one 18-hour day per

TABLE 2.4. *Nationality of the Workers at the State Oil Fields, 1917*

Nationality	No.	Pct.	Nationality	No.	Pct.
Spanish	324	23.1%	Argentine	46	3.3%
Portuguese	187	13.3	Italian	38	2.7
Russian	186	13.3	Bulgarian	28	2.0
Austrian	108	7.7	Other	342	24.4
Greek	91	6.5	TOTAL	1,401	99.9%
Rumanian	51	3.6			

SOURCE: *La Protesta*, Oct. 5, 1917, p. 3.
NOTE: Percentages do not add up to 100 owing to rounding.

week, without overtime, was obligatory). Skilled workers earned more, but averaged only 74 centavos an hour in 1919, often below what they could make for similar work in private industry.[54] These wages would not buy much at local stores, which charged three pesos for a dozen eggs, 90 centavos for a pound of sugar, and one and a half pesos for a head of white cabbage. Practically no milk or meat was available at any price, and high freight rates raised the prices charged for what was available at least 100 percent above prices in Buenos Aires.[55] Despite the completion of the aqueduct in 1913, water remained scarce and expensive at Comodoro Rivadavia. Ordinarily sold to the public at 50 centavos per 100 liters, water became almost unobtainable during droughts. On at least one occasion, frustrated townspeople protested a water shortage by smashing the government's tanks and letting the water flow down Comodoro Rivadavia's main street. Though the Chief Administrator (the man in charge of the Bureau's operations at Comodoro Rivadavia) proudly announced in 1919 that water was free for workers, in fact the labor force often could not obtain enough to drink. Moreover, there was never enough water to permit bathing, which added to the already serious hygiene problem. Visiting government agronomist Hugo Miatello emphasized that the water shortage could be solved, as considerable local resources existed, but the state failed to make the necessary investments until 1920, when Yrigoyen finally expropriated Behr's Estancia in order to ensure an adequate water supply for Comodoro Rivadavia at all seasons.[56]

Workers at Comodoro Rivadavia also had to put up with dismal housing conditions. The Petroleum Bureau had not constructed nearly enough housing to accommodate the influx of workers that began in 1914, and by 1917 the shortage had reached crisis proportions. Yet *La Prensa* reported that the Bureau had constructed three sumptuous mansions, "whose cost could not be less than 100,000 pesos each," for top managers, and that lower-level managers and supervisors, together with their families, occupied a variety of recently constructed houses and apartments.[57] However, single workers, the vast majority of the labor force, received totally inadequate housing—though they paid very low rents. About half the single workers lived two or three to a room in recently constructed sheet-iron barracks; the rest (over 500 in 1917) were housed in dirty, dilapidated shacks that had little or no heat and that served as havens for *bichucas*, parasitic insects endemic to the region whose bite caused purulent ulcers. Each room in these shacks, reported *La Prensa*, measured three by four meters and housed six workers who lived in "the worst filth." In the nearby camps of the private oil companies, on the other hand, workers lived in "well-constructed buildings with electric lights."[58] Medical facilities at Comodoro Rivadavia were totally inadequate before 1924. The one infirmary, which was "a shack built of zinc sheets," had a capacity of four beds.[59] Workers injured on the job found that the Petroleum Bureau paid no compensation for work-related accidents, though the national workman's compensation law of 1916 clearly was intended to cover laborers in this type of industry.[60]

These working conditions, along with poor labor-management relations, spawned severe unrest among the state oil workers that culminated in a major strike on September 29, 1917. Curiously, most Argentine oil historians pay little or no attention to this strike—it lasted 51 days—and the long series of work stoppages that followed it and that severely disrupted petroleum production between 1917 and 1920. The 1917 walkout began after a group of workers organized the Federación Obrera Petrolífera (FOP), a local union that initially received no funds or organizers from the outside.[61] The FOP requested an eight-hour day, a 25 percent raise for workers earning less than four pesos daily and a 15 percent raise for those making more than that, and overtime pay. The union took these demands to Leopoldo Sol, who had been named Chief Administrator of the oil field in 1911, when the Petroleum Bureau took over operations at Comodoro Rivadavia. When Sol rejected these proposals, almost the entire work force, including 45 drilling foremen, went out on strike. Production halted completely. In response, the government sent the battleship *Rivadavia* and the cruiser 9 *de Julio* to Comodoro Rivadavia, where they landed several hundred sailors ostensibly to protect the petroleum installations. After the strikers refused an order to return to work, management not only fired them but evicted them from their lodgings.[62] On the day of the evictions, the navy set up machine guns in the streets of Comodoro Rivadavia. Navy technicians attempted to resume production at some of the wells, but they succeeded only in starting destructive fires. Mechanics and firemen from the fleet also began to operate the state's Colonia Sarmiento railway, whose workers had struck in sympathy with the oil laborers.[63]

To try to present their case to the government, a workers' committee left Comodoro Rivadavia by car late in October and journeyed some 900 kilometers north to Río Colorado, where they entrained for Buenos Aires. On October 31 they met with Minister of Agriculture Pueyrredón. They reiterated their complaints, emphasizing that management's arbitrary and uncompromising attitude long had alienated the work force, and requested the navy's withdrawal. A few days later, however, Chief Administrator Sol arrived in the capital by ship to give his version of the dispute. He claimed that outside labor agitators were to blame and that "the best and quickest way to settle the strike was to take energetic measures against the leaders of the movement." Both Sol, who threatened to resign, and the Petroleum Bureau's Administrative Commission stated that "for technical and economic reasons" the workers' demands could not be met. One member of the Commission, Enrique Hermitte, pointed out that adoption of an eight-hour day would require three daily shifts instead of two, causing a labor shortage and raising wage costs at least 33 percent.[64]

Anxious to end the strike and resume oil production, Yrigoyen disregarded the hard-line advice of the Administrative Commission and charged Pueyrredón first to confer with Sol and two navy captains sent as observers to Comodoro Rivadavia and then to negotiate a settlement of the dispute on the basis of their testimony. After listening to the naval officers' analysis of the strike, which conflicted with Sol's, Pueyrredón reached an agreement with the workers' committee. On November 9, the government promised to raise wages by 20 percent for workers earning less than four pesos per day and by 10 percent for those earning more; to institute a 9.5-hour day (with provisions for overtime); to improve housing, medical care, and food supply; and to rehire all the strikers.[65] By November 18 the workers were back at their jobs, but the long stoppage had cost fuel-starved Argentina about 32,000 cubic meters of oil worth around two million pesos. Moreover, several important wells had become inoperative and remained closed for months after the strike.[66]

In protest against the President's conciliatory policy, Sol and the entire

Administrative Commission resigned. Yrigoyen used this occasion to placate the navy, whose interest in oil we have already noted, by appointing Captain Felipe Fliess the new Chief Administrator at Comodoro Rivadavia on November 12. The new Chief Administrator's principal staff assistants also were navy officers. Fliess, who had been assistant director of Argentina's naval academy, had made a detailed study of the state's oil fields and prepared a lengthy report for the navy hierarchy during the 1917 strike, but otherwise had no experience in the oil industry. Though Yrigoyen's reorganization did not place the oil fields under control of the Ministry of Marine, the naval command assumed that it would henceforth exercise strong influence over the industry. The Ministry itself stated that naval officers would "impose in the administration . . . a certain military character, necessary in treating this [industry, which is] so closely linked to the military needs of the fleet . . . and in which such a large group of workers of all nationalities with the most extreme tendencies is employed." [67]

But the navy's plans for a vigorous and efficient administration of the state oil fields came to naught, for Yrigoyen had no intention of granting a branch of the military power over an important national resource. In a fashion characteristic of his presidency, Yrigoyen dispersed authority at lower levels in his 1917 reorganization in order to centralize decision-making in his own hands. The President did not fill the vacant seats of the Administrative Commission and simply left the oil fields within the Ministry of Agriculture, which sent its own accountants and overseers to Comodoro Rivadavia. During the next five years, the Agriculture Ministry, underfinanced and pressed with other business, often neglected or ignored the oil fields' most elemental needs. Fliess, in his struggle to increase production, often came into conflict with Joaquín Spinelli, the Agriculture Ministry's chief petroleum accountant and the most powerful official at Comodoro Rivadavia because of his control of finances. Severe administrative confusion—and disputes between the Agriculture Ministry and Fliess —resulted from this dispersion of authority and plagued the state oil industry during the rest of Yrigoyen's first presidential term. [68]

Captain Fliess found his term as Chief Administrator at Comodoro Rivadavia a profound exercise in frustration. When he assumed his position, he announced plans to drill at least 125 new wells yearly; according to his projections, this would enable oil production to increase to 950,000 cubic meters by 1924. But Spinelli refused to release the necessary finances. Moreover, severe shortages of supplies and equipment, owing in part to the

4. Captain Felipe Fliess inspecting works at Comodoro Rivadavia, November 1918.

unwieldy administrative structure and in part to wartime shortages, obstructed Fliess's ambitious plans. The Chief Administrator struggled in vain to obtain an adequate supply of motors, drilling equipment, pumps, and pipe. Argentina's fledgling national metals industry was unable to remedy the lack of pipe, and pipe became so scarce that Fliess had to order workers to strip it from existing wells in order to provide it for newer ones, an action that put 16 producing wells out of service in 1919.[69] Primarily because of these shortages, the state oil fields saw only 30 new wells drilled in 1918, rather than the 40 originally planned; in the following year new wells drilled declined to 19.[70]

Most of the state's Comodoro Rivadavia reserve remained virtually unexplored throughout the war years as a result of the Ministry of Agriculture's decision to open a new oil field at Plaza Huincul in Neuquén Territory, where President Figueroa Alcorta had established a state reserve of suspected oil lands by a 1909 decree. Early in 1916, the Ministry began to concentrate much of its scarce drilling equipment at Plaza Huincul, but cold temperatures, lack of water, and physical isolation forced work to proceed slowly, and the first Neuquén well did not begin production until September 1918.[71] Though output from the Neuquén fields was low, the excellent quality of the crude oil aroused great interest in Buenos Aires, and the Ministry of Agriculture pursued exploration vigorously.[72]

After the 1917 strike and administrative reorganization, labor unrest continued to plague the Patagonian oil fields and to disrupt production. A brief strike occurred in mid-November 1918, after Fliess refused to consider a demand presented by the FOP for additional wage increases to compensate for the erosion of purchasing power caused by the rapid wartime inflation. The strike ended after about a week, when Fliess decreed a raise of 50 centavos per day; but later in November it resumed after he denounced five labor leaders as "agitators," fired them, and shipped them out of Patagonia on a naval transport.[73] The FOP, which viewed this action as inaugurating a period of militaristic union-busting, responded by totally paralyzing oil production. This strike began just as anti-Bolshevik paranoia was gripping the Argentine upper classes, who viewed the nation's increasing labor unrest as a revolutionary threat. During the Semana Trágica early in January 1919, the Yrigoyen government would demonstrate its determination to deal severely with the labor movement by bloodily repressing an anarchosyndicalist general strike. Fliess prefigured that development with his heavy-handed policy in the oil fields. After the cruiser *Garibaldi* landed 200 sailors to "guarantee order," Fliess threatened to fire and expel from Patagonia all laborers who failed to return to their jobs by December 30. To underline his determination, the Chief Administrator set navy men to work at 12 wells in an attempt to resume production. Fighting broke out between strikers and a group of strikebreakers protected by the navy, and several members of the FOP were wounded. These tactics broke the strike, but the labor force remained disgruntled, unrest continued, and production suffered as a result. The reason the government was able to take a harder line against striking oil workers in 1918 than in 1917 was that imported oil, Mexican in origin, was again becoming available to ease the fuel crisis.[74]

Congress and the Petroleum "Trusts"

The series of strikes and shortages that kept the state's oil production at disappointing levels during the first two years of Yrigoyen's term reflected the President's failure to formulate a decisive petroleum policy. Although Yrigoyen occasionally paid lip service to the needs of the oil industry, he avoided action other than the ill-fated administrative reorganization of 1917. Moreover, he prudently abstained from taking sides on a series of petroleum-policy initiatives that influential congressmen launched early in his presidency.

Again, the mixed-company idea surfaced. In June 1916, a group of Radical Party deputies headed by Tomás de Veyga of the Federal Capital had proposed legislation aimed at increasing oil production as rapidly as possible. De Veyga asked Congress to create a Sociedad Anónima Argentina de Petróleo—a company in which the state would hold controlling interest but in which private Argentine capitalists could invest—to operate the state's Comodoro Rivadavia fields. The plan, which resembled Demarchi's 1914 proposal discussed in Chapter 1, attracted support from Argentine capitalists, who were anxious both to increase the nation's fuel supply and to partake in the profits of production. But de Veyga's mixed-company proposal appeared to some congressmen to open the door to foreign control over the government's oil fields. After Yrigoyen took office in October, he gave the plan no support, and it died in committee.[75]

The 1916 congressional session also brought the divisive issues of federalism and provincial rights to the forefront of Argentine oil politics. Although exploration of the rich oil deposits in Mendoza, Salta, and Jujuy provinces was only beginning, foreign oil companies already were hastening to acquire concessions from the provincial governments (as we shall see in later chapters). Congressional supporters of the state oil enterprise—who were primarily from the petroleum-consuming regions of the country, and especially from the capital and Buenos Aires province—viewed these concessions as a matter of grave concern and a menace to the future of the state company. Moreover, since foreign companies had established a record in other countries of acquiring concessions to establish reserves rather than to produce oil, deputies from Argentina's cities feared the provinces' concession policies threatened the future availability of cheap fuel in the country's populous regions. Accordingly, in August 1916, Radical Deputy Carlos F. Melo of the Federal Capital and Conservative Deputy Rodolfo Moreno of Buenos Aires province introduced legislation to amend the 1886 mining code by placing all oil and coal concessions under the jurisdiction of the national government.

Moreno's argument that this amendment would keep the foreign "trusts" under control did not impress deputies from the interior, however, who viewed any proposal to reduce the provinces' sovereignty over their mineral resources as an attempt of the federal bureaucracy and the urban population to tap provincial wealth solely for their own benefit. For the most part poverty-stricken and underdeveloped, the Argentine interior provinces had stagnated economically for half a century while the coastal regions had prospered. Oil offered the interior provinces new hope of economic growth, and the elites there were convinced that the interior would fare better by dealing directly with private petroleum investors than by surrendering their oil concession rights to the national government. Francisco M. Uriburu, an influential Conservative lawyer from Salta who would represent Jersey Standard from 1919 to 1927, led the opposition to the Melo-Moreno proposal, which occupied the attention of the Chamber of Deputies during both the 1916 and the 1917 legislative sessions. In speech after speech, Uriburu insisted that the provinces' rights to administer petroleum concessions was inalienable. The Melo-Moreno proposal was unconstitutional, for "the dominion of the provinces over their resources . . . springs not from the law but from their territorial rights, from sovereignty superior . . . to national organization and to the constitution itself, which recognizes and guarantees it." Undoubtedly well aware that certain Radical deputies from the interior agreed with Uriburu, and determined not to divide his party, Yrigoyen remained silent during the debates and allowed the Chamber to send the matter to committee to die.[76] The federalism issue would assume paramount importance in Argentine petroleum affairs during the 1920's, and the congressional debates over the Melo-Moreno proposal clearly indicated that a national oil-concession policy would open deep political wounds.

Congress did move swiftly on one petroleum matter in 1917, though; that concerned regulation of the oil-refining business, almost all of which Jersey Standard controlled. The issue that led to action was the tariff on imported kerosene, then a vital item of mass consumption. In 1906, to protect the infant Compañía Nacional de Petróleos (CNP), founded the previous year, the government had imposed a tariff of three centavos (gold) per liter on imported refined petroleum products: gasoline, kerosene, and lubrication oils. Crude oil continued to be admitted duty free. As we have seen, in 1911 the CNP, with its Campana refinery, came under control of The West India Oil Company (WICO)—a Jersey Standard subsidiary —and WICO adopted the practice of importing partially refined products under the label of crude oil to avoid paying duty. To end this practice, and to do something to curtail the rapidly rising cost of living, Conservative Deputy Francisco J. Oliver of Buenos Aires province introduced a proposal in 1916 to halve the duty on kerosene, to impose a duty of one-half centavo (gold) per liter on crude imported for refining, and to continue to admit free of duty crude destined exclusively for use as fuel oil.[77]

When the Chamber of Deputies took up debate of the Oliver proposal in January 1917, the legislators' attention immediately focused on the power of the Standard Oil Company and its Argentine subsidiaries. A strong coalition of Socialists, Radicals, and Conservatives attacked the existing duty on kerosene as an unnecessary tax on Argentine consumers and as a source of high profits for the Campana refinery, now under foreign control. Victor M. Molina, a Radical deputy representing the Federal Capital, portrayed the Standard-owned refinery as a "pseudo-industry, which does not employ nationally produced raw materials." To underscore his criticism, Molina exclaimed that Standard had "amassed its fortune with the blood, the tribulations, and the civil war of a brother republic"—a reference to Standard's role in the Mexican Revolution.[78] Socialist deputy Nicolás Repetto joined Molina in supporting the tariff on crude destined for refineries, for such a tariff would end the process by which Standard's subsidiaries were "defrauding the treasury" by "introducing an impure liquid that is basically kerosene but that looks like natural petroleum." The Yrigoyen government, through Treasury Minister Salaberry, expressed its support for the tariff changes and claimed that the industry surely could continue to exist under the new import duties.[79] Argentine commercial interests, acting through a "National Chamber of Commerce," joined in the chorus supporting the changes and argued that the exemption from duty benefited Standard Oil alone.[80]

Standard's subsidiaries attempted to counter these arguments, but without success. A pamphlet issued by CNP warned that the proposed tariff reduction might force "the closing of the plant and the ruin of the industry, since the firm would not be able to operate profitably." Argentine investors in the firm voiced similar concerns during the congressional debates. But when the measure reached the Senate, one of the leaders of that house, Benito Villanueva of Mendoza, rejected the company's threat to close as a "bluff." A majority of both houses agreed, and the tariff changes became law on February 14, 1917.[81]

The congressional criticism against Standard Oil's refining and distributing activities begun during the kerosene tariff debates continued when the Chamber of Deputies launched an antitrust investigation in 1917. In an attempt to placate the public's furor over high prices and consumer-goods shortages, which had become two of Argentina's most pressing political questions during the First World War, the Chamber appointed a special committee, chaired by Deputy Juan B. Justo, leader of the Socialist Party, to study the power "trusts" held and the practices they employed. The "trusts" investigated with particular care included meat packing, sugar, and kerosene. The committee's final report, issued in 1919, generated great public interest throughout Argentina. It contained a severe indictment of Standard Oil and emphasized that WICO, which sold about seven-eighths of the kerosene consumed in Argentina, had monopolized the trade in that product. Testimony from witnesses revealed that WICO had seized control of the kerosene market by undercutting competition and by paying bonuses to wholesalers who agreed to sell no other brands. Once WICO took over the market, the Standard Oil subsidiary assured its continuing dominance by obliging wholesalers to buy its kerosene exclusively at whatever price it decided to fix. Retailers, moreover, had to sell at prearranged prices. Any violation of this procedure would cause WICO to stop its kerosene deliveries. Though the committee found that there was no proof the "trusts" had increased prices absolutely, it did find that WICO garnered excessive profits relative to the cost of production.[82]

The tariff debate of 1917 and the subsequent antitrust investigations also publicized the fact that WICO had gained control of about 80 percent of the nation's gasoline trade and enjoyed almost complete dominance in the capital city, by far Argentina's most important gasoline market. Moreover, the congressmen learned that a 1915 Buenos Aires municipal ordinance clearly had fostered Standard Oil's predominant position in the capital. This ordinance specified that gas pumps in the streets had to be located at least 400 meters apart; and since the firm of Guillermo Padilla, Ltda., which was owned by a major Argentine stockholder in CNP and purchased supplies exclusively from WICO, had already erected pumps throughout the city, particularly in the most lucrative locations, the ordinance effectively prohibited other firms from entering the gasoline retail trade to compete against Padilla and WICO.[83]

Convinced that Standard Oil and its subsidiaries enjoyed excessive power in the Argentine petroleum trade, the antitrust committee proposed a series of policy changes designed to foster competition and protect consumers. It was impossible, the investigators argued, to counter this "international and almost worldwide monopoly" with a "simple reduction of customs duties," because "every private enterprise will be annihilated or absorbed by the trust." To provide effective competition against WICO's monopoly, the committee urged the government to promote the vigorous development of the state oil industry, which in 1919, with the exception of a small refinery at Comodoro Rivadavia, still limited its activities to producing and selling crude. The antitrust investigators advised the government to set up a vertically integrated industry to refine and sell the oil it produced. Finally, the committee recommended antitrust legislation.[84] Eventually, in 1923, Congress passed laws prohibiting price-fixing, the imposition of fixed resale prices on buyers, and other acts tending to foster monopolies.

The Emerging Oil-Policy Conflict

The significance of the antitrust investigation and the earlier kerosene tariff debate was to isolate Jersey Standard politically from Argentina's consuming masses, most of whom by this time lived in the coastal cities. Acting in the name of consumers, the congressmen had charged Standard with a list of practices that kept prices high and destroyed competition. During the decade following the First World War, the state oil company not only would provide effective competition but would force the foreign companies to reduce prices significantly, thus generating strong political support in the cities and particularly among the urban middle class.

The war years also revealed that regionalism would block the national government's attempts to control the foreign oil companies in the provinces. One of Salta's members of the Chamber of Deputies had raised the banner of federalism and provincial rights to prevent enactment of a national petroleum code, and during the 1920's the oil-producing provinces repeatedly would support Jersey Standard and fight attempts to permit the state oil company to begin operations in the interior. The interior provinces thus became allied with the traditional enemies of the government petroleum industry, who long had argued that state capitalism was a disastrous precedent for the Argentine economy.

Argentine oil politics were developing into a conflict between the consuming regions, especially the cities, and the oil-rich interior provinces, and the first Yrigoyen government attempted to defuse this conflict by returning to the prewar economic order. Committed to promoting the export trade, Yrigoyen and his advisers viewed the economic disruption of the First World War not as a danger signal warning that the era of export-led growth was ending but as a temporary aberration. Accordingly, the first Radical government ignored the theories of industrial protectionism that Alejandro Bunge was formulating. Hopeful that the international trading system would function smoothly again and that Argentina would thrive as a beef and cereal exporter, Yrigoyen relegated development of industry, including the state oil fields, to a minor priority. Petroleum imports, it was assumed, again would provision the urban population efficiently. After the First World War, however, the international oil companies upset this political equation when they moved quickly into Argentina to acquire the best remaining oil lands. Though he hesitated to act decisively before 1922, the political consequences of this oil rush ultimately would convince Hipólito Yrigoyen to throw his full support behind the state oil industry.

CHAPTER III

The Postwar Oil Rush and the Birth of YPF,
1919-1922

President Yrigoyen's 1922 decision to reorganize the state oil company under the name of Yacimientos Petrolíferos Fiscales ("State Petroleum Deposits"; hereafter YPF) marked the end of a long period of government indifference toward the fate of the national petroleum enterprise. YPF's future appeared bleak in 1922. Administrative confusion, labor conflicts, and charges of scandal had demoralized the state petroleum company for several years. Moreover, it faced a growing challenge to its paramount position in Argentine oil production from the European and North American oil companies that rushed into Argentina and the rest of Latin America during the global search for petroleum resources following the First World War. Argentina's lack of national petroleum legislation enabled the companies to obtain rich concessions while the state firm still was limited to producing on its two original reservations. These events of the immediate postwar period pointed to the bitter conflict that would take place after 1922 between YPF and the foreign oil companies. Determined to secure new reserves to make possible the state enterprise's expansion, the leaders of YPF would view the concessions the foreign companies had obtained as a dangerous threat. The foreign oil companies, particularly Jersey Standard, would become the prime targets of Argentine petroleum nationalism. The strategy and tactics that the foreign companies employed to acquire oil concessions would convince petroleum nationalists that nothing less than a state monopoly would save Argentina's oil for YPF.

The Global Struggle for Oil

Anglo-American rivalry for control of the world's petroleum resources, along with ever-increasing demand in the industrialized countries, stimulated a major international oil rush during the early 1920's. In the United States, geologists gloomily pointed out that the nation, which produced 65 percent of world oil output, held about one-twelfth of known world reserves, enough for only 30 years at the prevailing rate of consumption.[1] Dr. George Otis Smith, Director of the United States Geological Survey, warned in 1920 that the country would have to reduce consumption or depend more on foreign sources—a dangerous gamble should another war break out. His comments and those of other officials fostered the mood of crisis. Meanwhile, British writers, proud of the fact that British and Dutch companies by 1919 had acquired over half the world's estimated reserves, alarmed the U.S. government and press. Particularly disturbing was an article entitled "Britain's Hold on the World's Oil," which Sir Edward Mackay Edgar, a prominent British oilman, published in September 1919. The article claimed that "the British position is impregnable" and that "it will be within the limits of the commanding position that the future has in store for us to hold up the entire world for ransom in the distribution and production of this vital essential." Unless the United States acquired major new sources, it seemed only a matter of a few years before the country would become tributary to British oil interests; some experts predicted that the oil import bill from Great Britain might rise as high as a billion dollars annually.[2]

In Britain, the government and the oil companies pressed forward in a determined campaign to strengthen the British position in the world petroleum industry. As the London *Times* put it, "there has . . . arisen a national conception of oil as power." Overt diplomatic support for British overseas oil companies had begun well before the war, for Britain was even then the world's third largest petroleum consumer and the Royal Navy was rapidly converting to oil fuel.[3] In 1913, under pressure from naval interests, the government acquired a majority interest in the Anglo-Persian Oil Company. After the war, Royal Dutch Shell, in which English investors held a large interest, greatly expanded its drilling, its tanker fleets, its refineries, and its marketing facilities worldwide. "We must not be outstripped in our struggle to obtain new territory," wrote Sir Henri Deterding, head of Royal Dutch Shell, in 1920. "Our interests are therefore being considerably extended, our geologists are everywhere any chance of success exists." The San Remo agreement of 1920 between Great Britain and France further fed American suspicions of impending British world oil supremacy. This pact, in which the British and French agreed to coordinate their oil policies for mutual advantage and to give France a share in exploitation of Iraq's oil, appeared to exclude U.S. companies from one of the Middle East's richest fields. Reports from the mandated territories of the Middle East indicated that the British were strengthening their position with laws and administrative regulations that effectively barred Americans from obtaining oil concessions.[4]

Faced by the rapid expansion of British oil companies, the U.S. State Department in 1919 launched a diplomatic campaign to aid American companies attempting to develop foreign sources of petroleum. Public policy, as Leonard Fanning put it, now assumed that oil supply had a "vital national defense complexion." The State Department advised all U.S. embassies and consulates to secure information on oil discoveries, concession grants, and changes of ownership of petroleum lands. In another circular, on October 7, 1919, the Department advised its field representatives that this information was "urgently needed." Moreover, Washington authorized its diplomats "to lend all legitimate aid" to Americans seeking oil concessions abroad for the present and future needs of the United States. The government attached particular importance to extending the Open Door principle to allow all oil companies—whatever their provenance—the opportunity to acquire petroleum concessions in newly discovered oil fields throughout the world. This was a policy the British Foreign Office rejected, however.[5] Backed by a sympathetic State Department, U.S. oil companies increased their total investments abroad from $400 million to $1.4 billion during the years 1919–29. Yet the British-Dutch group also rapidly increased its foreign investments, so that by 1930 American companies controlled a smaller proportion of foreign oil production than they had in 1918, whereas the British controlled about the same proportion of production. Despite the continued advance of the British oil companies, major new petroleum discoveries in the United States and elsewhere reduced British-American tension over oil by 1923. In that year London acknowledged the Open Door principle for petroleum investments, and American companies began to invest in Iraq and the Dutch East Indies.[6]

The worldwide scramble for oil concessions, which Myra Wilkins has termed a "massive panic," extended to Latin America during the early

1920's. As the decade opened, experts estimated that Latin America contained about a third of the world's total petroleum reserves, an estimate that strongly stimulated the interest of U.S. oil companies.[7] Jersey Standard in particular set out to expand its production of crude oil in Latin America, for in 1918 the company produced about 16 percent of the crude processed in its refineries; moreover, Jersey Standard faced political problems in the United States and severe British competition in the Middle East, and these factors undoubtedly contributed to its decision to expand in Latin America.[8] The rise of economic nationalism in Mexico, which had just undergone nearly a decade of revolution, made it risky to invest there, so Standard and other American companies focused their interest on the oil resources of the South American continent. During the 1920's, American oil investments in South America rose from $67 million to $476 million; the percentage South America represented in total U.S. oil investments abroad rose from 17 percent in 1919 to 34 percent in 1929. Stimulated by foreign investments, South American oil production rose from less than one percent of world output in 1919 to nearly 13 percent in 1930.[9]

Peru's rich petroleum deposits had attracted the interest of foreign oil companies late in the nineteenth century. The British-owned La Brea-Pariñas field near Talara on the northern coast produced the bulk of Peru's petroleum, and until the Venezuelan oil boom of the 1920's Peru was South America's leading oil exporter. The British were supplanted after the First World War, however, as dictator Augusto B. Leguía moved Peru firmly into the U.S. diplomatic and economic sphere: in 1924, after years of negotiations, a Jersey Standard subsidiary, the International Petroleum Company (IPC), acquired title to the La Brea-Pariñas fields; and Royal Dutch Shell was rebuffed in its attempt to obtain exclusive exploration rights in Peru.[10]

Jersey Standard also firmly established itself in Peru's neighbor Bolivia, in whose Santa Cruz region rich reserves were known to exist. The Standard Oil Company of Bolivia, organized in November 1921 as a subsidiary of Jersey Standard, first bought out some three million hectares of oil concessions that the U.S. prospecting firms of Richard Levering and of Spruille Braden had previously obtained, and then acquired an additional million hectares of Bolivia's best potential oil lands from the government. Standard's first Bolivian well began production in 1925; output was low through the rest of the decade, partly because of transport difficulties, but Standard's operations in Bolivia nonetheless aroused great concern in neighboring Argentina.[11]

Jersey Standard matched its success in the Andean republics by gaining a dominant position in the petroleum production of Colombia. An American firm, the Tropical Oil Company of Pittsburgh, had begun exploration and drilling there as early as 1916, and by 1919 had brought in several wells on its rich concession 64 kilometers southwest of Barranquilla on the northern Colombian coast. Jersey Standard acquired Tropical Oil in 1920 and rapidly expanded its activities. By 1927, Standard's output from its Colombian holdings exceeded that from any of its other foreign properties. Produced largely for export, Colombian oil flowed to coastal ports in pipelines owned by another Standard subsidiary. Though at least 20 other U.S. firms and ten British ones also entered Colombia to search for oil in the 1920's, none met with success.[12]

Though Peru, Bolivia, and Colombia were important to foreign oilmen, no South American country attracted as much interest as Venezuela. Unsuccessful in most of the rest of the continent, Royal Dutch Shell, which was reducing its Mexican operations, began significant production in Venezuela in 1917 and made major discoveries around Lake Maracaibo in 1922. Venezuela's rich deposits, low transport costs, and stable government under the iron-fisted rule of dictator Juan Vicente Gómez stimulated a major oil boom during the rest of the 1920's. To attract foreign oil companies, Gómez enacted a petroleum code in 1922 that oilmen lauded as "the best in South America."[13] Besides Royal Dutch Shell, Gulf and Standard of Indiana also expanded dramatically; the two U.S. companies, which had produced only .5 percent of Venezuela's output in 1922, produced 54.8 percent of it by 1929.[14] During the 1920's boom, Venezuela replaced the USSR as the world's second-largest oil producer and became the chief foreign supplier to the United States.[15]

Though the Venezuelan boom reduced the interest foreign oil companies displayed in most other South American countries, throughout the 1920's a number of British and American concerns competed to gain control over the petroleum resources of Argentina. As in the rest of South America, the Argentine oil rush was motivated partly by the international companies' desire to acquire oil reserves for the future. But the Argentine situation differed fundamentally from that in any of the continent's other oil-producing countries, for the domestic market for petroleum products in Argentina was large, lucrative, and rapidly growing. By 1920 Argentine consumption of petroleum products exceeded that of France, and the republic was becoming, in the words of oil expert J. A. Brady, "one of the great consumers of petroleum products in the world."[16] At the beginning of the 1920's, Argentina was the world's sixth-largest importer of refined gasoline and seventh-largest importer of lubricating oil.[17] Moreover, unlike Peru, Colombia, and Venezuela, Argentina was swept by automobile fever during the 1920's, with the result that by the end of the decade the republic ranked seventh in the world in motor vehicles registered (with 310,805— more than double the number registered in Brazil, which ranked second in South America). In response, Argentine demand for gasoline and other petroleum products grew voraciously during the 1920's; by the end of the decade the republic's gasoline consumption was about half the continent's total. Production of domestic petroleum to supply this expanding market became an attractive business proposition—particularly in the case of Jersey Standard—for the duty on imported crude oil the Argentine congress had levied in 1917 raised production costs at the company's Campana refinery.[18]

Argentina's Oil Rush

Oil policy had become a major political concern in Argentina by 1919, as we saw in the previous chapter, but the country still had no comprehensive national petroleum legislation and continued to regulate oil development under the terms of the 1886 mining code. As the March 1920 congressional elections approached, Yrigoyen sensed the political appeal of petroleum issues among Argentina's urban voters—the core of Radical Party strength. Aiming to control the provinces' petroleum concessions in the national interest, the President revived the substance of the 1916 Melo-Moreno proposal and strengthened it with additional provisions. In September 1919, he presented legislation to Congress based on the principle that the national government was the original owner of all petroleum deposits in the republic. Specifically, the legislation proposed (1) to remove administration of oil concessions from the provincial governments and to grant the national government power to supervise all exploration and concessions; (2) to prevent monopolization of oil resources by private companies by requiring written authorization from the president for the transfer of ownership of any petroleum concessions; (3) to allow the president to declare future government reserves whenever he might deem it necessary; and (4) to reorganize the state oil industry under a newly created Dirección General de Yacimientos Petrolíferos Fiscales in the Ministry of Agriculture.[19]

In his message to Congress accompanying the oil legislation, Yrigoyen attempted to reassure private investors that the government did not plan to establish a state petroleum monopoly. He emphasized that the existing state reserves at Comodoro Rivadavia and at Plaza Huincul in Neuquén were sufficient for the government-owned oil industry, and that "a vast and rich field of action will always remain for private enterprise in the oil zone."[20] The Argentine ambassador to the United States issued similar assurances in an address to the American Petroleum Institute.[21] Yet despite the President's stated commitment to private participation in Argentine oil production, the 1919 legislation would have vastly enhanced the national government's authority over Argentine petroleum affairs and would have enabled it to regulate, or for that matter exclude, foreign investments in the provinces.[22]

Yrigoyen's proposals encountered a frigid reception in Congress, however; despite the fact that the Radicals gained a majority in the Chamber of Deputies for the first time in the March 1920 congressional elections, Congress failed even to consider the president's proposed petroleum reform. The Radical Party remained divided over the petroleum question, and prominent deputies from the interior continued to oppose any plan to centralize jurisdiction over oil concessions in the hands of the national government. Yrigoyen repeatedly urged the legislators to act, but they pigeonholed his proposals and Argentina remained without a basic oil law. In this context, private oil companies suffered few restraints during the speculative land rush of the early 1920's.[23]

Private investors already had established a secure foothold in Argentine petroleum lands by the time Yrigoyen proposed his oil legislation. The province of Salta was one of their major theaters of operations (investment there is discussed further in Chapter 4); the Patagonian territories of Neuquén, Chubut, and Santa Cruz, jurisdiction over which rested in the national government's Bureau of Mines, was another. Under Sáenz Peña private investors gained few concessions in Patagonia, but under de la Plaza the number of hectares conceded grew (see Table 3.1). During the fuel crisis of the First World War, particularly after 1917 legislation eased cer-

TABLE 3.1. *Government Oil Land Concessions to Private Investors in the Argentine National Territories, 1911–22*
(Hectares)

Year	Hectares conceded	Year	Hectares conceded
1911	14,262	1918	66,950
1912	11,372	1919	53,218
1913	16,393	1920	95,475
1914	99,619	1921	69,500
1915	43,790	1922	45,618
1916	48,782	TOTAL	646,845
1917	81,866		

SOURCE: Argentina, [2], Sesiones ordinarias, vol. 3 (Aug. 17, 1927), p. 821.

tain restrictions in the 1886 mining code, the Bureau of Mines granted oil concessions rapidly, so that by the end of 1922 investors had acquired title to 646,845 hectares in the three Patagonian territories.[24]

Foreign companies typically worked "under cover" to acquire concessions, as the Sinclair Exploration Company informed the State Department in 1921. A prospector, serving as agent of an oil company, obtained exploration permits and did the preliminary work required by law. Once he acquired title to the claim and made the initial investments the mining code required, he did not have to operate an oil well or do any further work to hold the claim, but merely had to pay an annual tax of 100 pesos per claim. Under these conditions, agents of foreign companies acquired extensive contiguous blocks of land.[25] Free-lance prospectors also obtained claims for speculative purposes, with the sole object of selling them to the highest bidder later on. To hold their exploration permits for as long as possible without doing expensive drilling work, speculators established a pattern of requesting exemptions and delays from the Bureau of Mines. Underfinanced and swamped by thousands of requests for exploration permits and concessions, the Bureau usually ignored the stern time limits for exploration set in the 1886 mining code. As a U.S. diplomat reported, "There is the greatest inclination on the part of the Department of Mines to interpret the mining code as liberally as possible."[26]

The foreign oil companies initially focused their attention on the rich oil lands adjoining the government's 5,000-hectare Comodoro Rivadavia reserve. Argentina's first private oil company, Astra (Compañía Argentina de Petróleo, S.A.), obtained a 1,500-hectare concession and began production near Comodoro Rivadavia in 1916. Though Argentine investors had

formed Astra, German interests, associated with Deutsche Erdölgessel-
schaft and aiming to help Germany recover the oil position it had lost in
the war, acquired control of the company early in the 1920's. The company
enjoyed close relations with the Yrigoyen government, which purchased
Astra oil for the navy.[27]

The British railways followed the lead of Astra and began investing in
the Patagonian oil fields. Convinced that oil fuel was both more economical
and more dependable in supply terms than imported coal, three of the largest British railways, the *Sud*, the *Oeste*, and the *Pacífico*, decided in 1920
to switch to oil. The companies formed a consortium, the Compañía Ferro-
carrilera de Petróleo, that leased for 20 years a 1,500-hectare concession
adjoining the state reserve at Comodoro Rivadavia from its owner, the
Compañía Argentina de Comodoro Rivadavia, a private Argentine firm.
Seven wells, which produced 2,200 cubic meters of oil a month, already
existed on the concession.[28] Once the contract was signed, the consortium
began constructing a modern refinery with a 1,300-cubic-meter daily
capacity at Comodoro Rivadavia, and the railways rapidly converted their
locomotives to burn oil.[29]

While the British railways were establishing a secure lease over Argentine oil lands, in London a group of British investors headed by Lord Cowdray, the magnate of Anglo-Mexican Petroleum, was formulating ambitious plans to acquire control of the state's Comodoro Rivadavia reserve
and to export part of its production. Early in 1920, Cowdray and his
associates—who had formed the Whitehall Petroleum Corporation, Limited—made public a proposal they had already presented to President
Yrigoyen outlining the formation of a mixed company in which Whitehall
would invest £5 million (50 percent of the share capital) and to which the
Argentine government would contribute its 5,000-hectare Comodoro
Rivadavia reserve (the other 50 percent). Whitehall also promised to contribute its experience, credit, staff, and management. The resulting company would enjoy a monopoly of exploration, production, refining, and
transport on the state's reserve for 51 years, and would receive exemption
from taxes as well as permission to export petroleum. In return, the state
would receive 65 percent of the net profits.[30] The London *Times*, which had
learned of the proposal in December 1919, strongly supported it. After
informing its readers that the United Kingdom, because of "her immense
permanent investments," had "more interest than anyone else in making
Argentina self-supporting," the *Times* argued that the best results at

Comodoro Rivadavia would arise under a system in which the state protects "its rights under a system of royalties, rather than by direct official working." British interests would benefit also, in view of "the close proximity of the Falklands," which "doubtless would provide permanent and profitable customers for such surplus as Argentina may be able to market. . . . How necessary it is for the trade of the world that a base for the provisioning of vessels with liquid fuel be established at Port Stanley."[31]

After lengthy negotiations with Whitehall's representatives, however, Yrigoyen rejected the proposal. He had demanded power to prohibit exports and to fix selling prices, two conditions Whitehall would not accept.[32] Analyzing the defeat of the Whitehall contract, British Minister Macleay informed the Foreign Office that Yrigoyen had been "misled" by a "sort of misguided patriotism which, as in China, regards any concession of mineral rights to foreigners as an irretrievable surrender of a priceless national asset." Moreover, added Macleay in a bitter note, "vested interests," including "Senators and Deputies," who purchased state oil to "retail it, naturally at an enhanced price, to friends and clients," had influenced the President's decision.[33]

Yrigoyen's position regarding foreign petroleum investment was more complex than the disappointed British diplomat admitted. As the Whitehall affair revealed, Yrigoyen opposed alienation of the state's oil reserve; and though his 1919 legislative proposals demonstrated his determination to control oil concession policy throughout the entire nation, he certainly did not oppose "any concession of mineral rights to foreigners." Indeed, the government placed no obstacles in the path of the Anglo-Persian Oil Company and Royal Dutch Shell's subsidiary Diadema Argentina, both of which acquired Patagonian oil lands during the speculative boom of the early 1920's.[34]

Investors from the United States also moved quickly to acquire petroleum concessions in Argentina. But in contrast to Lord Cowdray's associates, who had negotiated directly with the Yrigoyen government to form a monopoly over the state's Patagonian reserves, at least some U.S. investors took a different approach. In 1920 and 1921, a U.S. company called the Bolivia-Argentine Exploration Corporation formulated an ambitious scheme to flood the Argentine market with low-cost Bolivian oil. The immediate aim was to undercut and then absorb the state oil firm, with domination of the Argentine petroleum industry the ultimate goal. The evidence is inconclusive, but it suggests that Bolivia-Argentine was a front for Jersey Standard. Whatever the case, the nature of the scheme reveals the almost unbridled atmosphere of speculation and intrigue that pervaded Argentine oil affairs during the early 1920's.

The central figure in Bolivia-Argentine's activities was one Durelle Gage, a U.S. promoter who recently had worked on behalf of American investors in major Colombian and Bolivian oil land transactions. Early in 1920, the U.S. embassy in Buenos Aires learned that Gage, who was vice-president of Bolivia-Argentine, was negotiating oil concession contracts with a number of provincial governors. Once the Whitehall deal fell through, the embassy informed the State Department that opportunity awaited U.S. oil interests in Argentina. Bolivia-Argentine seemed to have a head start in the field, and embassy officials indicated that they would not be averse to lending what aid they could to see that a U.S. firm profited from the failure of the British initiative. But there was one important condition for such aid: Bolivia-Argentine must prove it was not connected with Jersey Standard, for according to Consul Wiley, no Argentine government would consider granting any "national petroleum monopoly on a partnership basis" to a firm connected in any way with Standard Oil because of Standard's negative image in Argentina.[35]

To forestall the possibility of future embarrassment on this score, the State Department launched an inquiry and learned that the well-known American investor Spruille Braden was president of Bolivia-Argentine, which was headquartered in New York City. We have already seen that Braden and another U.S. investor, Richard Levering, had acquired some three million hectares of oil concessions in Bolivia. In March of 1920, Standard Oil of New Jersey opened negotiations to acquire these oil lands from Braden and Levering. Despite the show of negotiations, the State Department suspected that both men had been acting "merely as agents" for Standard, as Herbert Klein has put it.[36] Determined to learn whether Bolivia-Argentine was now serving as a front for Standard in Argentina, Secretary of State Colby wrote Braden to inquire whether any other company "directly or indirectly" was associated with Bolivia-Argentine. Braden replied in the negative.[37]

Meanwhile, Gage was continuing his negotiations with Argentine provincial governors. On September 10, 1920, he signed a contract with Governor José Camilo Crotto of Buenos Aires province granting Bolivia-Argentine the right to explore for oil, which experts suspected existed in the province's southernmost regions. The contract stipulated that whereas any deposits the company might find would be the "exclusive dominion and property" of Bolivia-Argentine, they would be exploited by a mixed company in which the province would have a half interest. One month later, Gage signed a similar agreement with Governor Horacio Carrillo of Jujuy province. Attempts to arrange deals along the same lines with the provinces of Córdoba and Salta were well under way.[38] In a conversation with Consul Wiley in July of 1920, Gage had frankly admitted his two-pronged approach: "With three or all four of the provinces named under our control," he remarked, "the [federal] government, almost perforce, will be obliged to enter into a national scheme for exploiting the petroleum in this country." In the event that the national government resisted this notion, Gage intended to bring it to its knees by flooding the Argentine market with cheap Bolivian oil piped from southeastern Bolivia through "friendly" Argentine provinces to a Paraná River port and thence shipped by tanker to the Buenos Aires market.[39]

The State Department was not at all pleased with what it was learning either about Mr. Gage's South American activities or, on pursuing its U.S. investigation, about Bolivia-Argentine's. Washington was concerned that Gage's scheme might discredit other U.S. oil companies in future negotiations. But more important was the fact that there had been an apparent change of policy in the State Department toward U.S. oil companies doing business overseas. Washington, engaged in delicate negotiations with the British to achieve an "Open Door" policy for American oil companies in the Middle East, could hardly support the establishment of an American-controlled petroleum monopoly in Argentina. Signaling this change of attitude, Colby emphasized in a dispatch to the Buenos Aires embassy that "the Department does not look with favor on national or even provincial petroleum monopolies, but desires that its nationals should be given equal opportunity with others. The Department cannot support absolute monopoly pipelines concessions."[40] Answering the State Department's request for further information early in 1921, Spruille Braden claimed that he and his father, the wealthy developer of a large Chilean copper mine, were not financially involved in Bolivia-Argentine but served the company "mainly as advisory engineers." Financial backing for Bolivia-Argentine, Braden revealed, came from the Philadelphia banking firm of George W. Kendrick & Company—a firm that had backed Gage during his earlier Colombian and Bolivian oil land deals. The Kendrick firm, Braden continued, owned the company's Argentine properties, "although they are held in the name of Mr. Gage." On February 4, 1921, the State Department asked George Kendrick III to reveal his plans for the Argentine concessions.[41]

At this point, Kendrick abandoned Gage. Braden explained to the State Department that "financial conditions in the United States" dictated the decision to end operations in Argentina, but the Department's investigation into the relationships among Gage, Braden, and Jersey Standard may also have helped convince Kendrick to withdraw his support. With its funding cut off, Gage's scheme crumbled. Gage closed his office, and neither the Buenos Aires nor the Jujuy legislature ratified the contracts he had negotiated in 1920.[42]

Whether or not Jersey Standard was associated with the Gage scheme, the giant U.S. corporation demonstrated strong interest in beginning oil production in Argentina, and, surprisingly, did not find the Yrigoyen government uncooperative. Yrigoyen's position with regard to Jersey Standard, the only U.S. oil company that had significant investments in Argentina at this period, reflected his commitment to private investment but also his determination to regulate that investment in the national interest. Yrigoyen had urged Congress to pass his 1919 petroleum proposal in order to prevent private concerns from monopolizing oil production in the provinces, a development that he realized could lead to the establishment of a foreign oil monopoly over the entire country. But he also wanted to increase Argentine oil production, and he was willing to deal with foreign companies, including Jersey Standard, interested in investing in the national territories.

In August 1920, Jersey Standard sent a representative, Alberto A. Moreno, to Buenos Aires to approach the Yrigoyen government about oil concessions in Patagonia. On October 14, 1920, Moreno had an interview with Minister of Agriculture Alfredo Demarchi, the industrialist who had earlier expressed his interest in the development of Argentina's petroleum resources by proposing a "mixed-company" model in 1914. Demarchi, who had succeeded Pueyrredón in the Agriculture Ministry in 1918, remarked: "Personally, I realize that the fuel problems of this country are so serious that they cannot be solved by the government alone or by the native capitalists, who lack the necessary experience; hence I see no reason why your entrance into this field would not be welcome, if you adjust yourselves to the laws of the country." Demarchi agreed to give Moreno "all available geological information" so Moreno could send it to Jersey Standard "to guide them in determining what areas are available for private enterprise." When Moreno suggested that Standard considered Argentina's oil concession laws excessively vague, Demarchi responded cryptically that "lawyers will always find a way of adapting the laws to the requirements of the times." "If you wish to apply for exploration zones of 2,000 hectares on government lands," the minister concluded, "the matter will receive my favorable consideration; but I cannot at present offer you any facilities beyond those provided by the code and precedent." [43]

Apparently assured by Demarchi's remarks, Standard soon opened a Buenos Aires office and began to acquire oil concessions in Neuquén and Chubut territories and in Salta province. After spending over two million pesos to acquire 10,000 hectares at Challecó, in Neuquén, the company formally incorporated in 1922 as The Standard Oil Company of Argentina, S.A.[44] The new, wholly owned subsidiary of Jersey Standard operated independently of WICO, Jersey Standard's other major Argentine company. Standard of Argentina began drilling in 1922 but did not make its first strike until 1924.[45]

State Oil Scandals and the Founding of YPF

While foreign oil companies were moving into Argentina to obtain concessions and begin production, the state oil workings were languishing and, through their inefficiency, were seriously discrediting the whole concept of a government-owned oil company. The administration of the state oil fields between 1919 and 1922 was little improved over the near-chaotic situation that had prevailed during the First World War. Mismanagement, financial shortages, and labor disputes continued, contributing a gloomy outlook for the future of the state firm. Moreover, Argentina remained dependent on oil imports, as Table 3.2 shows.

In 1919, the government unveiled a comprehensive improvement plan for Comodoro Rivadavia that envisioned an expansion of drilling and construction of better port facilities, new storage tanks, a hospital, and a freezing plant for meat.[46] But adequate financing to carry out these projects never became available, and the improvements progressed slowly, if at all. The development of the port languished, the tanker fleet remained inadequate, and oil continued to be stored on the ground until 1923.[47] True, 79 new wells were drilled between 1920 and 1922, and production did increase significantly.[48] But the private companies' impressive growth

5. Oil being loaded for transport, Comodoro Rivadavia, April 1921.

6. The port of Comodoro Rivadavia "in full activity," as the original caption put it, in February 1919. This was before the wharf was built.

TABLE 3.2. *Argentine Domestic Production and Imports of Petroleum,*
1914–22
(Cubic meters)

Year	Total petroleum consumption	Domestic production		Imports	
		Amount	Percent of total consumption	Amount	Percent of total consumption
1914	278,795	43,795	15.7%	235,000	84.3%
1915	541,580	81,580	15.1	460,000	84.9
1916	502,551	137,551	27.3	365,000	72.6
1917	547,374	192,374	35.1	355,000	64.9
1918	399,867	214,867	53.7	185,000	46.3
1919	751,300	211,300	28.1	540,000	71.9
1920	1,057,495	262,495	24.5	795,000	75.2
1921	1,296,905	326,905	25.2	970,000	74.8
1922	1,495,498	455,498	30.5	1,040,000	69.5

SOURCES: Dorfman, p. 143; "Resumen estadístico de la economía argentina," *Revista de Economía Argentina*, 20 (Nov. 1938), p. 326.

overshadowed the performance of the state firm—between 1919 and 1922 they expanded output about 400 percent and enlarged their share of total Argentine oil production from 11 to 23.4 percent (see Table 3.3).

The labor disputes that had plagued the state oil fields during the war continued to disrupt production in 1919 and 1920. On assuming command at Comodoro Rivadavia, Fliess had attempted to alleviate some of the most glaring labor abuses. He also did everything possible to see to the construction of adequate housing for the workers—though he had to admit to the Ministry of Agriculture in 1920 that at existing construction rates, unsatisfactory housing conditions would persist at least until 1922. In any event, outside observers charged that even the new housing was unhygienic and overcrowded.[49] In another attempt to improve conditions for the workers, Fliess established a commissary where they could buy provisions for 10 to 20 percent less than they could in town. But a reporter from the Buenos Aires newspaper *La Razón* noted in 1919 that the commissary functioned "with no benefit to the workers, for the merchandise is sold there at the same prices as in the private shops."[50]

The inefficacy of Fliess's measures, coupled with Argentina's general postwar economic distress, produced a highly volatile labor situation in the oil fields, where strikes became epidemic late in 1919 and early in 1920. In mid-August 1919, members of the FOP, the oil-workers' union, walked off the job and forced production to halt. Though Fliess claimed that "noxious elements" and agitators caused the strike, serious disputes over work-rule changes imposed by newly hired overseers appear to have been the most important cause. As in 1918, Fliess resorted to force, arbitrarily closing union meetings and giving the leaders one-way tickets to Buenos Aires. The strike did not end until mid-September and resulted in an estimated loss of 25,000 cubic meters of petroleum production.[51] By this time, Fliess was requiring all overseers and foremen to join the Liga Patriótica Argentina, a right-wing organization that had been formed to combat strikes in the name of fighting Communism.[52]

Another, more serious strike over wages began on December 21, 1919, and lasted more than two months. Severe inflation had made the wage increases of the past two years nugatory; however, the drop in the market price of fuel after the war left the government less able to give pay increases than in previous years. Moreover, as imported oil flowed freely into Argentina and eased the fuel crisis, the government could afford to take a much more intransigent attitude against striking oil workers than it had in 1917 and 1918. Hence in December 1919, when the FOP asked for a 20-percent wage increase (a request that would have cost the government an extra 70,000 pesos a month), Fliess countered with an offer of about 10 percent. When the FOP rejected his offer, 1,700 workers, nearly the entire labor force at Comodoro Rivadavia, left their jobs.[53]

Fliess adopted a hard line in dealing with the strikers. Refusing to recognize the FOP as the workers' bargaining agent, the Chief Administrator

TABLE 3.3. *State and Private Petroleum Production in Argentina, 1914–22*
(Cubic meters)

Year	Total domestic production	State production				Private production	
		Chubut	Neuquén	Total	Pct. of total domestic production	Chubut	Pct. of total domestic production
1914	43,795	43,795	—	43,795	100.0%	—	—
1915	81,580	81,580	—	81,580	100.0	—	—
1916	137,551	129,780	—	129,780	94.4	7,771	5.6%
1917	192,374	181,704	—	181,704	94.5	10,670	5.5
1918	214,867	197,573	13	197,586	92.0	17,281	8.0
1919	211,300	188,093	19	188,112	89.0	23,188	11.0
1920	262,495	226,544	612	227,156	86.5	35,339	13.5
1921	326,906	277,807	919	278,726	85.3	48,180	14.7
1922	455,498	343,910	4,978	348,888	76.6	106,610	23.4

SOURCE: "Resumen estadístico de la economía argentina," *Revista de Economía Argentina*, 20 (Nov. 1938), p. 323.

arrested union leaders and again called in navy troops, who broke up union meetings and demonstrations and expelled strikers and their families from their housing at bayonet point. Joining in the repression were local members of the Liga Patriótica Argentina, whose leader was the manager of the Comodoro Rivadavia branch of the Banco de la Nación. Fliess announced that he was willing to consider granting wage increases over the 10 percent proposed, but only on condition that he personally select the work force. When a delegation of workers appealed directly to Minister of Agriculture Demarchi, he told them curtly that state workers were forbidden to strike. The Buenos Aires establishment press supported this policy of intransigence, and *La Prensa*, which blamed the walkout on "agitators," criticized Fliess for negotiating with the workers at all.[54]

In order to break the strike, the government's Immigration Bureau began to recruit new oil workers, particularly experienced Germans and Alsatians, to go to the Patagonian fields. The Buenos Aires anarchosyndicalist newspaper *La Protesta* published telegrams from the FOP appealing to the working classes to resist this policy of strikebreaking. But the FOP's appeals were in vain, for the Germans and Alsatians began arriving at Comodoro Rivadavia in mid-January, and by the end of February the government had succeeded in breaking the strike.[55] Rather than accept the failure of their strike, hundreds of workers instead accepted a government offer to convey any workers who wished to leave to Buenos Aires. Indeed, a full 90 percent of the labor force at Comodoro Rivadavia left their jobs during the course of 1920—*and* 53 percent left soon after the strike. While Fliess was superintending this wholesale turnover of the work force, production was gradually resuming and approached prestrike levels in mid-March. The government's reliance on strong-arm tactics to break the strike had temporarily destroyed the organized labor movement at Comodoro Rivadavia—but at the cost of seriously disrupting production.[56]

The strikes and interruptions of production at the Comodoro Rivadavia oil fields stirred opposition members of the Chamber of Deputies to launch an investigation of the government's petroleum policy in August 1920. Nicolás Repetto, an outspoken Socialist deputy from the Federal Capital, opened the attack on the government in an address to the chamber on August 24. Charging that despite party rhetoric the Radical administration had done little since 1916 to develop the state oil fields, he introduced a resolution to invite the Minister of Agriculture to report to the Chamber on the state of oil production, transport, and sales, and on future plans

7. An oil worker at Comodoro Rivadavia, January 1919.

for the Comodoro Rivadavia fields. After bitter opposition from Radical delegates, the Chamber voted to approve Repetto's resolution on September 2.[57] But the Yrigoyen government continued its policy of slighting congressional requests for information, and the Minister of Agriculture refused to appear. Instead, Yrigoyen sent the deputies a few pages of barren statistics that purported to review state oil operations between 1916 and 1920. The President's attitude enraged the opposition, which responded by accusing the government of corruption and condemning its petroleum policy. Repetto charged that Comodoro Rivadavia was in "total chaos, in a state of permanent conflict and strikes," and asked why the government was using its tankers to transport the oil of a private company, Astra, for six or seven pesos per ton while Astra's own ships were busy exporting oil at 30 pesos a ton. Both Repetto and other opposition deputies accused the government of restricting state oil production in order to aid private companies, particularly Astra, in which Minister of the Treasury Salaberry's firm Salaberry y Bercetche was known to own shares.[58]

The congressional inquiries of 1920 did succeed in revealing that labor unrest was only partly to blame for lagging production at Comodoro Rivadavia, for it came to light that the state was deliberately holding down output. In its yearly report for 1920 the Petroleum Bureau admitted it had cut production, but claimed that rapidly falling prices of both imported oil and coal were creating a saturated oil market and competition that the state oil fields could not meet.[59] What it did not admit, though, was the languorous fashion in which it responded to this competition. For after importers had been cutting prices for months, the Ministry of Agriculture finally decreed reductions of the high wartime prices of state-produced oil in late October 1920. Because of the Ministry's sluggish reaction, importers were able to increase their share of the Argentine market in 1920 by 250,000 cubic meters and the government could only sell 68 percent of what little it did produce.[60] By early 1921, when state production exceeded storage capacities, output was actually declining; by mid-1922, only a third of the government's wells were in production, the rest having been temporarily capped. Yet during this period the neighboring private oil fields were working at full capacity. The government's policies of restricting production while watching passively as oil imports boomed and foreign oil companies rushed into Argentina set off sharp protests in the press in 1922. Both La Prensa and La Vanguardia attacked the government for negligence in failing to provide adequate storage and transport, which would have allowed production to increase greatly.[61]

Rumors of scandal swept Buenos Aires as the state oil industry deteriorated, and they gained credence when Fliess resigned his post as chief administrator in August 1921 to protest the government's administrative confusion. Though Fliess was replaced by another navy man, retired Captain Francisco Borges, real power at Comodoro Rivadavia fell to Joaquín Spinelli, whose control over finances as the Ministry of Agriculture's chief accountant had constantly frustrated Fliess.[62] During 1922 and 1923, investigators from the press and from Congress learned that the Ministry of Agriculture, acting through Spinelli, had for years failed to formulate coherent budgets for the oil fields, supervising finances simply through arbitrary decrees. Because the oil fields had no budget for the year 1922, for example, the state had hired hundreds of unnecessary workers and administrators. Accounting was in miserable shape, and often there were no records of costs or sales.[63] The almost total absence of control over expenditures was paralleled by the rankest lack of concern for the actual needs of the oil fields. "Not even the most indispensable materials were available," an oil laborer of this period later recalled. "We didn't even have ropes. . . . It was true disorder."[64] Accusations of corruption also enveloped the Bureau of Mines, whose Director, the respected engineer Enrique Hermitte, resigned in January 1922 after Yrigoyen insisted on appointing untrained and technically inexperienced men to the Bureau's staff.[65]

The impending scandal broke at last in May 1922. Minister of Agriculture Demarchi had resigned in March of that year, and Yrigoyen's new appointee, Emilio Vargas Gómez, soon uncovered evidence that corruption and favoritism pervaded operations at Comodoro Rivadavia and that there were severe irregularities in the sale of state-produced oil. Vargas Gómez brought his preliminary findings on the petroleum situation before the cabinet, but the other ministers, at the behest of Yrigoyen, refused his request for a full investigation. Assuming an attitude rare for a member of Yrigoyen's cabinet, Vargas Gómez protested: he refused to authorize payments for the state oil fields' debts and, alleging ill health, failed to go to his office or to use his official automobile. It was at this point that the press, intrigued by this conflict between the Minister of Agriculture and the President, began to investigate the oil situation in earnest and to expose corruption.[66] Faced with the embarrassing situation his own minister had exposed, Yrigoyen decreed an administrative reorganization of the state oil industry on June 3, 1922. The Petroleum Bureau was abolished, and in its place was created a new Dirección General de Yacimientos Petrolíferos Fiscales to operate the government's oil industry. Still under the Ministry of Agriculture, the new agency was ordered to purchase materials through public tender and to recognize all debts incurred by its predecessor. Shortly afterwards, in August, Vargas Gómez resigned.[67]

Much of the Argentine press viewed the President's latest petroleum decree with skepticism. The *Buenos Aires Herald* doubted that the new structure would work better than the old one, which it dubbed a "rotten" and "ragtime oil administration." The "men at headquarters," commented the *Herald*, "have been just blundering through, making no effort whatsoever to support their colleagues in the south." As a result the staff sometimes went for months without pay, and "when the fields have called for money for this or that they have been in the position of a wife asking her liverish husband the price of a new hat." In agreement, *La Prensa* interpreted the creation of YPF as a move "to pass the sponge over the record of the oil administration," and predicted that the new organizational scheme, because it subordinated YPF to the Ministry of Agriculture, would continue to encourage corruption. The oil administration, it went on, should be "entirely independent in a commercial form."[68]

The scandal over the state oil industry soon reverberated in Congress,

8. "The Extraction of Government Oil," cartoon from *La Vanguardia*, July 23, 1922. The cartoon refers to the oil scandals of 1922 that led to the formation of YPF. The crowd is apparently composed of government favorites, Radical politicians, etc. all hoping to get their share.

where a group of deputies headed by Conservative Rodolfo Moreno introduced a resolution on July 19, 1922, asking the Minister of Agriculture to appear before the Chamber to respond to 57 pointed questions about the oil fields, the payroll, and the rate of production. Emphasizing that petroleum "was the most important economic question presently facing the country," Moreno, who long had urged a nationalist oil policy, charged that the government had deliberately deceived the country by blaming the low levels of production on strikes or lack of transport.[69] Radical deputies rejected Moreno's charges and countered with arguments that production demonstrated a long-term upward trend. When the roll-call vote on Moreno's resolution was taken on July 20, the Radicals prevailed 60 to 50 despite the defection of 12 of their number.[70] *La Prensa* condemned the Chamber's decision not to force a confrontation with Yrigoyen. In a sharply worded editorial, it emphasized that the president's party was trying to conceal corruption and that the executive branch was flouting the consti-tutional rights of Congress by refusing to provide information necessary for it to formulate intelligent oil legislation.[71]

To soothe criticism that it was concealing malfeasance, the Radical leadership in the Chamber introduced a weak proposal that invited the President to supply general operating information on the oil industry and that did not include the embarrassing questions Moreno had posed. The Chamber approved this proposal on July 28, after having defeated an amendment sponsored by Nicolás Repetto that asked Yrigoyen to explain the organization of the oil administration between 1916 and 1921 and to furnish a complete list of all new concessions granted to private oil com-panies during that period.[72] Nevertheless, Yrigoyen never answered even this mild request for information that his own party sponsored. It was clear that Yrigoyen adamantly opposed congressional investigation of his oil policy, and that the vast majority of Radical deputies remained loyal to him on an issue they condemned as an opposition-inspired political maneuver.

After Yrigoyen left office, clear examples of corruption in his govern-ment's oil policy emerged. In 1923, the Chamber of Deputies published the results of a lengthy investigation that substantiated many of the charges made when the oil scandal had erupted the previous July.[73] Also in 1923, the federal courts uncovered evidence clearly demonstrating that Yrigoyen's treasury minister, Domingo E. Salaberry, had worked with Standard Oil's subsidiary WICO to evade import regulations and defraud the treasury. Accusations of corruption had colored Salaberry's career for years: in 1921, for example, a group of deputies had accused him of favoring friends and his own business firm when issuing permits for the export of wheat, sugar, and scrap metal, and of condoning irregularities in the sale of state-owned land and in the adminstration of the customs houses. A motion to censure Salaberry failed by only four votes in the Chamber in April 1921. U.S. Consul White considered Salaberry the "most astute" member of Yri-goyen's cabinet and commented that "his finesse can best be judged by the fact that he has firmly convinced the President that he is an honest man." But in 1923, without the powerful support of Yrigoyen as president to sustain him, and in the face of the mounting evidence against him, Sala-berry committed suicide shortly after the federal court at La Plata impli-cated him in the oil import fraud case.[74]

The oil import fraud case began early in 1923, when Angel Guillermo

Frontini, a private citizen who had invested in Comodoro Rivadavia oil, brought a lawsuit against Jersey Standard's subsidiary WICO and its subsidiary, the Compañía Nacional de Petróleos (CNP). The suit, which incoming President Marcelo T. de Alvear's Ministry of the Treasury joined as party plaintiff, charged the companies with deliberately defrauding the national treasury by evading payment of customs duties stipulated in the 1917 oil tariff law. This legislation, discussed in Chapter 2, placed a duty on imported crude oil destined for refining but admitted crude oil destined exclusively for use as fuel free. The state presented evidence that WICO had imported partially refined oil for resale, and crude for refining, under the label of crude destined for fuel. Acting on Salaberry's advice, WICO evaded those provisions of the 1917 law that required identification of the origin of imported oil by using vague identifying labels.[75] Moreover, Salaberry permitted the company to falsify the minimum inflammability temperature that the law required to be posted on all petroleum imports. Documents presented to the court showed that when a cargo of oil labeled "USA Southern States" arrived at the Campana Customs House in 1917, Salaberry instructed the Chief of Customs that "the declaration of origin or derivation that the company makes will be enough." This practice prevailed from 1917 until Salaberry left office in 1922.[76]

The court convicted the Standard Oil subsidiaries of customs fraud and ordered them to pay the government the sums they had evaded, but the companies appealed the decision and secured a reversal in 1924 from a federal court of appeals.[77] The government then took the case to the National Supreme Court, which in 1928 upheld the court of appeals, absolving the companies on the grounds that they were following the Treasury Minister's instructions.[78] Regardless of the decision, the case conclusively demonstrated that during Yrigoyen's first presidency high officials of the Customs Bureau and the Minister of the Treasury himself had illegally favored foreign-owned oil-importing companies.

Yrigoyen's Petroleum Policy: An Assessment

While Yrigoyen, who constitutionally could not stand for reelection, was preparing to leave office in 1922, *La Prensa* examined his petroleum policy and found much to criticize. The patriotic duty of all Argentines, the newspaper editorialized, was to combat foreign oil companies and to prevent them from dominating the market. But, it continued, Yrigoyen's failure to promote the state oil fields had encouraged petroleum "trusts" to consolidate their position in Argentina.[79] Examination of the government's petro-leum policy between 1916 and 1922 partially supports this conclusion. By 1922, the reputation of the state oil industry had sunk to the lowest point in its history—the industry had virtually lost credibility as a serious business enterprise. Meanwhile, Jersey Standard, Royal Dutch Shell, and Anglo-Persian were expanding their Argentine operations rapidly. The oil importers still effectively controlled the Argentine market. *La Prensa*'s criticism of Yrigoyen's administration, however, overlooked the President's pioneering attempt in 1919 to construct a truly national petroleum policy that would centralize jurisdiction over provincial oil concessions in the hands of the federal government. The national jurisdiction plan, which Congress had refused to consider, would have enabled the Yrigoyen government to regulate the foreign oil companies in, or exclude them from, precisely those regions of the country where they were now maneuvering to establish monopolies over petroleum production.

Foreign investors, including the oil companies, expected to enjoy a favorable political atmosphere during the incoming presidential administration of Radical Marcelo T. de Alvear. Of distinguished family and cosmopolitan background, Alvear inspired "full confidence" among the diplomatic community in Buenos Aires.[80] But the oil companies were in for a surprise, for Alvear backed an extremely aggressive petroleum policy. He not only moved sharply to restrict the foreign oil companies; he also revitalized YPF by placing it under Colonel Enrique Mosconi, a dedicated officer who viewed development of the state oil industry as a matter of the highest national priority. Under Alvear and Mosconi, YPF would emerge as the very symbol of Argentine economic independence.

CHAPTER IV

Alvear, Mosconi, and the Rise of YPF, 1922-1927

The Argentine army became a powerful ally of YPF during the 1920's and furnished the dynamic, enterprising leadership that revitalized the moribund state oil company. Convinced that the country required economic change and that they possessed the technical skills necessary to lead that change, influential army officers began to advocate the industrialization of Argentina. These "entrepreneur generals," as sociologist José Luis de Imaz has called them,[1] recognized that petroleum was vital to the self-sufficient, industrialized Argentina they envisioned, and opted for statist solutions to develop the republic's oil industry. That the military technocrats enjoyed the resolute support of President Alvear from 1922 to 1928 casts doubt on the traditional interpretation of Alvear as a do-nothing conservative. To head the state oil company, Alvear appointed Colonel (later General) Enrique Mosconi, whose career became virtually synonymous with the rise of YPF. Mosconi not only promoted the rapid expansion of YPF but also reorganized it into a vertically integrated enterprise that could compete successfully against the international oil companies. During the late 1920's, Mosconi would draw upon his experience as Director-General of YPF to formulate an ideology of petroleum nationalism that has strongly influenced Argentine economic policy to the present day. Mosconi's ideology, along with his dramatic leadership of the state oil company, made YPF a symbol of Argentine economic independence.

The Alvear Presidency and the Radical Party Schism

Alvear's presidency coincided with a return of export-led prosperity to Argentina. Following a sharp postwar recession, demand for Argentina's rural exports began to rise steadily in 1924, and remained strong through 1928. In the latter year, the value of exports exceeded one billion gold pesos for the first time since 1920. New foreign investment, insignificant since 1913, again flowed into the country, and Argentina experienced another economic boom. Late in 1927, the republic returned to the gold standard for the first time since 1914.[2] Despite this prosperity, Argentina's position in the international economy remained precarious. As Chapter 5 analyzes in detail, most of the new investment came from the United States, which also replaced Great Britain as the main supplier of Argentine imports. Yet the Coolidge government enforced restrictions against the importing of meat into the United States that virtually closed the country to this key Argentine export and served to strengthen Argentina's export dependence on Britain, still the principal market for the products of the pampas.

It was in this context that Alvear supported Mosconi's plans to strengthen YPF and to challenge the foreign oil companies—particularly Jersey Standard, which was expanding its Argentine operations rapidly. Such a policy was not a threat to Britain: once the immediate postwar oil rush had tapered off, the British government showed itself more concerned with retaining its coal and oil export trade to Argentina than with promoting British petroleum production there. Its links with the British secure, the Alvear government had little to lose in terms of Argentina's international economic position when it mounted basically an anti-American petroleum policy.

Though Argentine relations with Washington experienced severe strain during the Alvear presidency, symbolic expressions of friendship between London and Buenos Aires underscored the determination of the British and Argentine governments to maintain their close economic ties. Before he took office in 1922, Alvear made a state visit to London, where George V entertained him at Buckingham Palace and invested him with the Grand Cross of the British Empire. In 1925, Alvear reciprocated with a lavish welcome when the Prince of Wales made a state visit to Argentina. Although Congress refused the President's proposal to appropriate 400,000 pesos for the occasion, Alvear nonetheless staged one of the most elaborate public celebrations in Argentine history and entertained the Prince with full-dress parades, fireworks displays, polo matches, and concerts at that monument of Argentina's age of splendor, the Teatro Colón.[3]

This close association with royalty was typical of Alvear's career. The new President was a member of the traditional upper class, segments of which had long been prominent in the Radical Party's leadership. Born into one of the republic's most aristocratic families in 1868, Alvear graduated from the Faculty of Law at the University of Buenos Aires in 1891. While a student, he acquired a liberal political ideology, became a member of the Unión Cívica Radical, and took part in the party's abortive 1890 revolution. But during his youth neither the legal profession nor politics absorbed Alvear's primary attention. An enthusiastic devotee of both sports and the arts, this sophisticated *porteño** traveled extensively in Europe and developed a strong affection and admiration for France. After marrying a Portuguese opera singer in 1906, Alvear settled in a villa near Paris, a city he loved "passionately," as his biographer Felix Luna has put it. There he became an intimate associate of some of France's most distinguished leaders, including Clemenceau and Poincaré.[4]

Returning to Buenos Aires in 1912, Alvear was elected both to the Chamber of Deputies on the Radical ticket and to the presidency of the Jockey Club, Argentina's most exclusive social organization. Yrigoyen offered him the Ministry of War in 1916, but Alvear declined and accepted instead the post of ambassador to France, where he remained until 1922, largely out of touch with domestic Argentine politics. Although Alvear disagreed with Yrigoyen's foreign policy, which kept Argentina out of the League of Nations, the Radical chief nonetheless selected him as the party's candidate for the 1922 presidential election, hoping thereby to maintain his influence over the government by passing over more prominent Radical aspirants.[5]

Yrigoyen kept control of the party machine, but his plans to dominate the government even while out of office received a setback as soon as Alvear assumed the presidency. Determined to govern through men of his own choice, Alvear quickly overturned a number of key appointments Yrigoyen had made late in his term. Moreover, the new President adopted a policy of financial stringency that contrasted sharply with the free-spending ways of his predecessor and reduced the patronage available to the Radical

* A *porteño* is a resident of the city of Buenos Aires.

9. President Marcelo T. de Alvear, official portrait.

10. President Alvear and Edward, Prince of Wales, on the latter's arrival in Buenos Aires, August 1925.

machine. The results of these changes, which had the backing of the party's elite sectors, were both a bitter political dispute between Alvear and Yrigoyen and a growing schism in the Radical Party.[6] Throughout its history an uncertain alliance of heterogeneous groups, the Unión Cívica Radical in the mid-1920's divided into two hostile factions, the "personalists," who followed Yrigoyen and desired his return to power in 1928, and the "antipersonalists," who hoped to obstruct that return by sponsoring federal intervention in key Yrigoyen strongholds.

The schism reflected not only a conflict over Yrigoyen's role but also class and regional divisions within the Radical movement. The party's political machine, centered in Buenos Aires city and province and linked closely to the masses and the middle-class bureaucracy, followed the personalists, whereas the party's aristocratic elements and much of its leadership in the interior provinces followed the antipersonalists. Although Yrigoyen during his presidency had attempted to avoid taking positions on economic questions that might weaken Radical support in the interior provinces, wartime and postwar inflation forced him to make decisions that fostered regional unrest within the party. In 1920, during a bitter conflict over sugar export and price policy between producing interests from the northwestern provinces and spokesmen for the urban consuming masses, Yrigoyen at first vacillated but eventually acted to keep prices down; this decision deeply antagonized the sugar-dependent provinces and drove some northerners, notably Deputy Benjamín Villafañe, out of the party. Elected governor of Jujuy in 1924, Villafañe became a leader of the antipersonalist faction and one of the nation's most vocal critics of Yrigoyen and his policies.[7]

Although he was the titular leader of the antipersonalist movement, Alvear played a hesitant and vacillating role that impeded efforts of the more militant leaders of the faction to make antipersonalism a true national political movement with a popular base. On the one hand, the President appointed leading antipersonalists to his cabinet. (One of them, Minister of the Interior Vicente C. Gallo, made a definitive political break with the Yrigoyenists in August 1924 by leading his faction to constitute itself formally as a separate party, the Unión Cívica Radical Antipersonalista.) But on the other hand, Alvear both failed to grant the antipersonalists the financial backing they needed to create a patronage base for the movement, and refused to sponsor provincial interventions aimed against the Yrigoyenists. Alvear, in other words, looked forward to the eventual reunification of the party.[8] His economic policies, moreover, did little to attract firm support for antipersonalism in the interior provinces. The strongest base of the faction's support was in the littoral province of Santa Fe, and throughout his presidency Alvear carried out an economic policy that primarily benefited the urban areas and the export-oriented pampas provinces —the traditional core of Radical strength. He opposed a modest tariff increase that Congress passed in 1923, and he utilized executive decrees to prevent the emergence of industries he considered uneconomical. He took no decisive action to aid the ailing sugar industry. Most importantly, Alvear promoted a highly nationalistic oil policy that directly challenged provincial autonomy on petroleum matters.

The 1924 Radical schism cost the party its majority in the Chamber of Deputies, where personalist representation declined sharply (from 90 seats in 1923 to 61 seats in 1927). Meanwhile, the antipersonalists aligned with Conservatives, Socialists, and Progressive Democrats to carry out open political warfare against the Yrigoyenist Radicals. Beset by partisan conflict and by the President's refusal to give strong leadership to the antipersonalist cause, Congress became little more than a group of warring factions and deteriorated into impotence. To prevent action on the part of the "contubernio," as the personalists dubbed the coalition of their enemies, Yrigoyenist Radicals adopted tactics of political obstructionism and frequently boycotted sessions of the Chamber in order to prevent a quorum.[9] By 1925, the most urgent national business, including enactment of the annual budget, went unattended. When Congress did meet, it spent its time in "futile floods of oratory," as the London *Times* noted. Like the *Times*, most observers of the Argentine political process were appalled at the virtual collapse of Congress. *The Review of the River Plate*, for example, condemned the "utter disregard for the public interest" that the "barren and undignified spectacle" of Congress represented.[10]

In sharp contrast to Yrigoyen, Alvear appointed professionally competent cabinet ministers he could trust to rationalize the government's departments and to end the corruption, political favoritism, and administrative confusion that had characterized the Yrigoyen presidency. To head the Ministry of Agriculture, still responsible for the state's oil industry, Alvear named Tomás A. Le Bretón, Yrigoyen's widely respected ambassador to the United States. A man of great energy and ability, Le Bretón proved a most efficient administrator; for the first time since 1916, the Agriculture Ministry functioned effectively. Le Bretón possessed a clear appreciation of the state oil industry's significance for Argentine economic development, and emerged as the strongest backer of YPF in Alvear's cabinet.[11] Named Minister of War was General Agustín P. Justo, a professional civil engineer who had served as Director of the Colegio Militar for the previous 18 years. Justo, who also believed strongly in developing the state oil industry, brought to Alvear's attention an impressive officer, Colonel Enrique Mosconi, who at that time headed the Army Aeronautic Service. The President considered putting Mosconi's known organizing ability to work by naming him Minister of Public Works, but on Justo's advice he named him instead Director-General and President of the Administrative Commission of YPF.[12]

Economic Nationalism and the Army

Alvear made his decision to place an army officer at the helm of YPF at a time when discontent with the republic's traditional economic system was gathering momentum within the Argentine military. Despite the generally conservative stance most army officers maintained on social issues in the 1920's, a small but growing group—many of whose members came from the Engineering Corps—began to advocate industrialization and greater national self-sufficiency as essential foundations for Argentina's military security and economic future. As in many developing countries, economic nationalists in the Argentine army became concerned "with the role of the nation as a military unit" and became convinced that "the capacity to wage war depends on the presence of efficient industrial capacity," as Harry Johnson has put it.[13]

Argentina's wartime economic crisis was certainly a major force behind this military reexamination of economic issues, but it was the enunciation of nationalist political and economic theories by Argentine intellectuals during the 1920's that provided the framework of analysis for the military's dissatisfaction with the country's economic situation. Particularly persuasive were the ideas of Leopoldo Lugones, one of Argentina's best-known intellectuals of the period. During the 1920's, this famed poet and essayist turned a critical eye on Argentine society; in a 1924 speech that was often quoted, Lugones proclaimed that "the hour of the sword has arrived, for the good of the world." His hierarchical and conservative social theories reflected contemporary European right-wing thought, especially that of Charles Maurras, but his economic ideas closely followed those of Alejandro

11. Nationalist intellectual Leopoldo Lugones delivering an address ("De la soberanía a la potencia" [From Sovereignty to Power]) from the stage of the Teatro Opera in Buenos Aires, 1928.

Bunge. Decrying Argentina's "colonial economy," Lugones declared that the country must aspire to a new future, one of national power, industrial strength, and economic independence. Only the army, Lugones was convinced, could lead the necessary transformation. The professional journal *Revista Militar* quoted Lugones's economic ideas approvingly as early as 1927, and his prestige among army officers continued strong as the 1930 coup approached.[14]

The army's economic nationalists soon went beyond Bunge and Lugones —who argued that the government should limit its economic role to providing tariff protection and an infrastructure base to stimulate private industrialists—and became convinced that the state itself (with the aid of the army) should intervene to help finance and operate at least such vital industrial installations as aircraft factories and petroleum refineries, plus such further installations as the private sector proved unwilling or unable to back. Moreover, whereas Bunge and other precursors of economic national-

ism had limited their critique of the Argentine economy to its export
dependence and had not attacked the position of foreign capital in Argentina, some of the army's leading economic nationalists, including Mosconi
and Colonel (later General) Alonso Baldrich, strongly criticized the power
of foreign "trusts" and argued that key industrial sectors should be financed
by Argentine capital.[15] Baldrich later summarized this position, which he
consistently held during the 1920's, when he emphasized that "true nationalism not only consists in combating anarchy and communism in order
to ensure that the order and hierarchy essential to liberty prevail, . . . it
implies the duty of constant struggle against, and implacable opposition
to, . . . the trusts that monopolize the sources of wealth in order to submit
to their dominion those peoples so incautious as to admit them."[16]

In brief, the basic themes of modern Argentine economic nationalism—
including industrialization, economic self-sufficiency, hostility to foreign
capital, and the need for an activist state to promote publicly owned enterprise—appeared within the Argentine army during the postwar decade. In
place of an economy dominated by the traditional agricultural export trade,
which they thought condemned Argentina to the "inferior rank" of countries, the army's economic nationalists foresaw an industrialized, militarily
powerful nation whose economic strength would be based on development
of its largely untouched mineral and energy resources.[17] Minister of War
Justo set industrial growth as an official military goal, expanded the army's
own manufacturing installations, and followed a policy of provisioning
the army from national industry wherever possible.[18]

This policy represented a start, but several influential officers remained
critical of the government's failure to support mining, metallurgy, and
heavy industry. For example, Colonel Luis A. Vicat, a metallurgical engineer and a frequent contributor to Revista Militar, argued in its pages in
1926: "Our economic organization is absurd. Should an enemy blockade
the republic's ports, they would defeat us without firing a shot." In order
to escape this dependent and strategically dangerous economic situation,
Vicat emphasized that the state must promote a national iron and steel
industry.[19] Other officers urged the government to promote industrialization by providing credit, establishing apprenticeship programs, and raising
tariffs. The example of Italy was frequently evoked to support the conten-
tion that Argentina's apparent lack of high-grade iron ore need not prevent
the development of a domestic steel industry, and Alvear's policy of permitting the unrestricted export of scrap metal received scathing criticism.[20]

Argentina's experiences during the First World War made petroleum a matter of fundamental concern among the army's economic nationalists; and the connection between military power and petroleum resources, one of the war's principal strategic lessons, provided a favorite theme for military writers during the 1920's. "The horrible specter of fuel shortages afflicted all our industries," recalled Major Raúl Barrera, who made a strong plea for state-supported oil development.[21] Colonel Vicat, in a series of articles published in *Revista Militar* in 1923 and 1924, echoed similar themes: "We can all remember the colossal rise in all fuel prices during the First World War," and the "delays of our railways, which had to burn green wood." These experiences, he continued, clearly indicated that Argentina would be helpless in a future war unless it began to develop its energy resources at once. Although Vicat was not totally opposed to private petroleum investment, he advised the state to exercise "severe vigilance" and "rigorous control" over the oil industry.[22]

A few army officers moved beyond Vicat's cautious position on private oil investment and urged a YPF monopoly over the Argentine petroleum industry. One of the most outspoken and influential military figures to adopt the monopolist position was Colonel Baldrich. A professional engineer, Baldrich gained personal experience in the petroleum industry between 1922 and 1924 as Chief Administrator of the Comodoro Rivadavia oil fields. During this period he became a zealous economic nationalist and supporter of YPF. A vigorous speaker and a prolific writer, Baldrich devoted much of his career to a campaign to convince the Argentine public that the state oil industry was essential to promote Argentine industrial development, to prevent foreign domination over the republic's petroleum, and to guarantee national military security.[23] Baldrich drew many of these ideas from Camilo Barcía Trelles, a Spanish university professor who wrote a book entitled *El imperialismo del petróleo y la paz mundial* in 1925. Widely read and much quoted in Argentina, this work argued that the Great Powers were engaged in a worldwide struggle to gain control over raw materials, especially oil, in the less developed countries. Mexico, according to Barcía Trelles, furnished a prime example of how the petroleum companies intervened in national politics to gain their ends. U.S. oil companies, he claimed, had openly taken sides during the Mexican revolution of 1910–17 and had granted strong financial support to friendly generals while opposing President Venustiano Carranza at every step. Concluding his wide-ranging historical survey, Barcía Trelles argued that the inter-

national oil companies were an "enemy" that it was "necessary to combat
. . . for the good of the world."[24] Baldrich commented that Barcía Trelles's
work "ought to serve as a text in our universities. . . . I recommend it
to the youth and to all uncontaminated men of my fatherland."[25] Promoted
to brigadier general and head of the Army Engineer Corps in 1926, Baldrich intensified his campaign against foreign oil investment in Argentina
during the late 1920's (see Chapter 5).

It is noteworthy—in view of the attitudes we have seen developing in
the army during the 1920's—that petroleum nationalism did not gain
significant support in the navy during the same period. Perhaps the confusion surrounding the state oil fields during the First World War convinced naval officers that a government-owned-and-operated petroleum
enterprise was not likely to prove a successful experiment. Certainly the
British training of the naval elite contributed to the greater prevalence of
traditional liberal economic thought among naval officers than among army
officers, many of whom looked to Germany for their model. Whatever the
reason, the navy appears to have favored the (British-model) mixed-company plan for developing the state oil fields once the wartime fuel crisis
was over. (See, for example, the discussion later in this chapter of Admiral
Lagos's proposals in his 1924 book *El petróleo en América*.) But the navy was
somewhat out of favor politically during the Alvear presidency—indeed,
Alvear's most powerful cabinet member was probably Minister of War
Justo—and thus its voice on oil policy was not influential.[26]

Mosconi at the Helm of YPF

Of the new group of "military captains of industry" that arose as a result
of the postwar ideological ferment pervading the Argentine army, none was
better known than Enrique Mosconi, who headed YPF between 1922 and
1930 and who continued to champion the cause of the state oil industry
until his death in 1940. Though he has been likened by some observers to
Rockefeller or Deterding, a more apt comparison is with Enrico Mattei,
head of the Italian state oil corporation ENI after the Second World War.
Like Mattei, Mosconi singlemindedly devoted his impressive organizational and administrative talents to the goals of strengthening and expanding the state oil industry and reducing the power of the foreign petroleum
companies. His efforts have made him a hero to successive generations of
Argentine petroleum nationalists.[27]

Enrique Mosconi was born in Buenos Aires in 1877, the son of an Italian

12. Colonel Enrique Mosconi, 1918.

immigrant engineer and a *porteña* mother. Like the sons of numerous middle-class immigrants, Mosconi chose the military as a career. Ambitious and hard-driving, he entered the Colegio Militar in 1891, graduated at the top of his class in 1894, and then began advanced engineering studies. After receiving a degree in Civil Engineering from the University of Buenos Aires in 1903, Mosconi gained field experience in projects with the State Railways and various government topographic commissions. As a captain, he spent the years 1906–9 in Germany as one of a small group of Argentine officers selected for advanced training there. Attached to the German army, he studied military engineering. "In that atmosphere of discipline and . . . of consummate technical-military experience," he later wrote, "we perfected our concepts of military service; there we learned devotion to duty, disinterest, professional honor."[28] During his stay in Germany, Mosconi first witnessed military aviation, which sparked an interest that endured throughout his career. Mosconi's ties in Europe became strong, for he spent a total of three more years in Germany, France, and Austria-Hungary in the course of two missions to purchase equipment for the army between December 1909 and December 1914.[29]

Promoted to colonel on his return to Argentina after the last of these two missions, Mosconi served as Subdirector of the army's War Arsenals between 1914 and 1919. In that post he became acutely aware that dependence on imported equipment jeopardized the Argentine military and began to emphasize the need for industrialization to correct the situation. In a 1918 speech to his comrades at the Arsenal, Mosconi lamented: "Our guns today are not independent; . . . we cannot provision ourselves sufficiently."[30] A pioneer of Argentine military aviation, Mosconi became Director of the Army Aeronautic Service in 1921. In this post he strongly promoted both military and civil aviation and began to put a national air communications system into effect.[31] Mosconi's aeronautical experience stimulated his interest in the petroleum question, for he learned to his distress that Argentina had to rely entirely on imported aviation fuel. As though that were not enough, WICO, which supplied the fuel, deeply offended Mosconi by breaking commercial precedent and refusing to sell to the army unless it paid in advance. It was this experience, Mosconi later recalled, that led him to swear "to struggle by all legal means to break the trusts."[32]

Alvear's appointment of Mosconi to head YPF was greeted with favor in the army, where the 45-year-old colonel's campaign for military industries and aviation earned him both popularity and prestige. (Mosconi was promoted to brigadier general in 1925, and served as president of the Círculo Militar, the Argentine army's most prestigious officers' club, during 1927.)[33] Throughout the 1920's, this man "who abounded with health and energy" fascinated foreign observers. As the Buenos Aires *Standard* described him, Mosconi was a "live wire" who had " no time for ornamental people who stand around and do nothing." U.S. Consul J. W. Riddle agreed that Mosconi was "capable and honest," but feared that he was "not very friendly to the United States." According to Riddle and other diplomats, Mosconi, who spoke French and German fluently but spoke little or no English, believed that Argentina's interests linked the country to Europe rather than to the United States.[34] In fact, the new head of YPF was primarily a nationalist who displayed an intransigent position during the 1920's against Argentine economic dependence on any country.

Mosconi approached his new post with a fervor that the Argentine state oil industry previously had not experienced. To his fellow workers he emphasized that a job with YPF was no ordinary position: identifying the state oil enterprise with Argentine patriotism, he portrayed each of its workers and employees as a civilian-soldier. His statement that "we all must completely dedicate ourselves—our minds, our hearts, and our muscle—to our work" summed up his determination to revitalize the state operation.[35] The new Director-General needed all the enthusiasm he could muster, for he took over a chaotic situation: the accounting system was backward; there were few reliable statistics; the administrative staff was disorganized and inefficient; no effective sales or distribution organization existed; ocean transport remained seriously deficient; the cost of petroleum production was too high; the technical personnel were inexperienced and listless; and the workers were demoralized. All these factors, Mosconi pointed out in a 1922 interview, restricted exploration and kept production at low levels.[36]

Mosconi devoted his first few months as Director-General to thorough study, both of YPF and of the Argentine petroleum trade. Convinced that YPF had to expand the scope of its operations to compete successfully with the private oil companies, by early 1923 Mosconi launched an ambitious reorganization and development plan. His goal was to create a vertically integrated state enterprise that would not only produce oil but refine it and market the resulting products at competitive prices.

The first step in Mosconi's plan was to remove YPF from the Ministry

of Agriculture's political interference: "The original organization scheme,
with authority divided between the Director-General and the Minister,"
and the resulting "lack of centralized management," he wrote, were "the
basic causes of all the difficulties . . . that the state operations experienced."[37] President Alvear responded sympathetically to Mosconi's reorganization plan and issued a decree on April 13, 1923, giving YPF complete administrative autonomy though leaving it technically part of the
Agriculture Ministry. Under the terms of Alvear's decree, the President
was vested with the right to appoint a six-man Administrative Commission
and a Director-General to operate YPF. The Commission was to decide
all policy matters by a majority vote, with the Director-General, who as
chief operating executive presided over Commission sessions, empowered
to vote only to break a tie. Under the new organization, YPF required
ministerial approval only for its annual budget and for major purchases
requiring the use of credit; otherwise, the Administrative Commission controlled all policy matters.[38] Mosconi considered the "ample autonomy" of
this organizational scheme one of the basic factors in YPF's later success.[39]

Once he had secured administrative autonomy, Mosconi prepared a four-
year plan to guide YPF's development between 1924 and 1927. The plan,
summarized in Table 4.1, projected an increase in crude oil production at
Comodoro Rivadavia from 493,000 cubic meters in 1924 to 1,943,000
cubic meters in 1927, and envisioned an expansion of the range of refined
products YPF placed on the market. Mosconi proposed to reach this ambitious level of output primarily through vastly expanded drilling in those
portions of the Comodoro Rivadavia fields already explored. He expected
to increase the number of producing wells there from 235 to 617 between
1924 and 1927. Assuming that production increases would earn profits
totaling 65.1 million pesos over the four-year period, the plan proposed
major new investments in YPF's capital plant[40]: new docks, waterworks,
and repair shops; a modern electric power plant to produce the electricity
needed to use efficient pumping machines in the wells; additional tankers
for ocean transport of oil between Comodoro Rivadavia and the urban
markets; and, most important, a major refinery.[41] The refinery, which
Mosconi predicted would be "the most solid pillar of our new organization," was the largest item in the Director-General's capital improvement
plan and vital to his vision of a vertically integrated petroleum industry.
YPF's existing refining capacity—limited to small plants at Comodoro
Rivadavia and Plaza Huincul—was totally inadequate and forced the state

TABLE 4.1. *YPF's Development Plan for Comodoro Rivadavia: Projections and Results, 1924−27*

Year	New wells completed		Wells in production		Crude oil production (m³)	
	Projected	Actual	Projected	Actual	Projected	Actual
1924	75	55	235	175	493,000	544,367
1925	100	129	312	257	862,000	609,272
1926	150	141	439	353	1,339,000	720,243
1927	200	127	617	470	1,943,000	792,364

SOURCES: Argentina, [24], foldout facing p. 64; Argentina, [34], p. 213; "Resumen estadístico de la economía argentina," *Revista de Economía Argentina*, 20 (Nov. 1938), p. 323.

enterprise to market most of its products in the form of crude oil. Without a large refinery, YPF lost the lucrative profits of refining and distribution, which remained firmly in the hands of foreign companies.[42]

Mosconi's plan enjoyed the strong support of President Alvear. In a message to Congress of June 12, 1923, the President proposed legislation to appropriate 20 million pesos to finance construction of a refinery for YPF. Congress rejected this allocation, so Alvear mounted a search for foreign contractors willing to build the refinery on credit. On December 31, 1923, the government approved a preliminary contract with the Bethlehem Steel Corporation of the United States. On the same day, with the full approval of his ministers, Alvear issued a decree authorizing YPF to finance the refinery through credit and loaning it 14.2 million pesos in Treasury Notes for preliminary payments to Bethlehem. The Treasury Notes were marketed with some difficulty, and only after Dr. Carlos Madariaga, a member of the Administrative Commission, guaranteed them with his personal fortune. Bethlehem, which agreed to accept payment in installments through 1926, began preparations for construction in 1924; the Alvear government ensured the refinery's completion when it extended YPF additional Treasury Notes totaling 10.7 million pesos in 1925.[43]

The Argentine government's decision to build its own refinery caused consternation among the foreign oil companies. Anglo-Persian, which had hoped to build and operate a refinery in a joint venture with YPF, believed that graft had played a vital role in influencing the government's decision. Sir Arnold Wilson, an Anglo-Persian executive, was reported to have said that Bethlehem got the contract through "the wholesale distribution of largesse to various highly placed Argentines, including the President and the Minister of Agriculture." Wilson also let it be known that he believed that Bethlehem was working as a front for Jersey Standard, which he

13. A symbol of Argentine nationalism: YPF's La Plata refinery (partial view, 1934).

thought would seize control of the entire Argentine oil industry once YPF proved incapable of running it. To the despair of British diplomats, these and later injudicious statements won Wilson no friends in the Argentine government, and in 1925 he left the country in virtual disgrace. British oilmen apparently were unaware that Jersey Standard officials were equally dismayed at the prospect of a YPF refinery, which they interpreted as an Anglo-Persian plot.[44]

In consultation with Bethlehem, YPF decided to locate the refinery near La Plata, capital of the province of Buenos Aires. The site was well chosen, for it adjoined the port of Ensenada, at which tankers from Patagonia could dock, and was only 48 kilometers from the country's largest market, the metropolis of Buenos Aires. When the La Plata plant began production in December 1925, it was one of the world's ten largest oil refineries in terms of capacity. Bethlehem operated the new plant for a year in order to enable YPF personnel to learn the business,[45] and during that first year of production, 1926, YPF's output of gasoline increased almost tenfold to nearly equal production at all privately owned Argentine refineries, of which Jersey Standard's at Campana was the largest. During the late 1920's, however, the privately owned refineries also increased their capacity and production (see Table 4.2). Despite its impressive first year, YPF's new refinery encountered serious technical difficulties owing to the fact that Comodoro Rivadavia oil was heavy in paraffin, which left a high proportion of residues. Consequently, in 1927 YPF signed another contract with Bethlehem for modifications in the original design and for a new cracking plant designed to increase gasoline yield by breaking down the residues of primary distillation. The new installation, completed in 1929, made possible significant increases in the La Plata refinery's output.[46]

Once the refinery was in operation, YPF was able to expand its product line. By 1922 the state enterprise had already begun to sell natural gas, which the Comodoro Rivadavia fields produced in abundance but which previously had been burned off. When Mosconi assumed control, he gave the highest priority to producing aviation gasoline, sale of which began in mid-1923. By the mid-1920's, YPF marketed a wide range of products, including kerosene, gasoline, and "Agricol," a fuel specifically designed for farm tractors. YPF diesel fuel appeared on the market for the first time in 1928.[47] Mosconi also attacked the difficult problem of creating a distribution and sales network to market YPF's increasingly wide range of petroleum products. As early as 1923 the enterprise began to set up the physical base for a national distribution system—storage depots in major cities to supplement its one existing depot, built at the Buenos Aires docks in 1914. The first of the new facilities, located at Rosario, Argentina's second-largest

TABLE 4.2. *Gasoline: Argentine Domestic Production and Imports, 1922–30*
(Cubic meters)

Year	Total gasoline consumption	Domestic production				Imports	
		YPF	Percent of total consumption	Private	Percent of total consumption	Amount	Percent of total consumption
1922	181,962	1,118	0.6%	22,323	12.3%	158,451	87.1%
1923	213,998	2,357	1.1	34,452	16.1	177,189	82.8
1924	286,949	4,689	1.6	45,093	15.7	237,167	82.7
1925	400,808	5,849	1.5	47,440	11.8	347,519	86.7
1926	524,250	48,525	9.2	56,991	10.9	418,734	79.9
1927	562,386	75,784	13.5	118,129	21.0	368,473	65.5
1928	693,812	99,212	14.3	195,811	28.2	398,789	57.5
1929	947,558	140,168	14.8	382,001	40.3	425,389	44.9
1930	961,795	178,650	18.6	442,998	46.1	340,147	35.4

SOURCE: Guevara Labal, *El petróleo*, pp. 93–94.

TABLE 4.3. *Argentine Domestic Production and Imports of*
Petroleum, 1922 – 30
(Thousands of cubic meters)

Year	Total petroleum consumption	Domestic production[a]		Imports[b]	
		Amount	Percent of total consumption	Amount	Percent of total consumption
1922	1,495	455	30.5%	1,040	69.5%
1923	1,720	530	30.8	1,190	69.2
1924	2,031	741	36.5	1,290	63.5
1925	1,802	952	52.8	850	47.2
1926	2,348	1,248	53.2	1,100	46.8
1927	2,772	1,372	49.5	1,400	50.5
1928	3,142	1,442	45.9	1,700	54.1
1929	3,393	1,493	44.0	1,900	56.0
1930	3,431	1,431	41.7	2,000	58.3

SOURCES: Dorfman, p. 143; "Resumen estadístico de la economía argentina," *Revista de Economía Argentina*, 20 (Nov. 1938), pp. 323, 328.
[a] Crude petroleum only.
[b] All petroleum products.

city, opened in 1923. Depots at Santa Fe and Concepción del Uruguay (distribution center for Entre Ríos province) followed in 1926, and at Mar del Plata and Ingeniero White (the port of Bahía Blanca) in 1927.[48] But YPF also needed a retail sales network. As late as 1925 it had only one gasoline sales agency in the entire republic, and associates of importing companies and the private oil refineries totally controlled sales of petroleum products. To rectify this situation and allow YPF products to be marketed as quickly as possible, Mosconi in 1925 signed a three-year retail contract with Auger y Compañía, a well-established firm with outlets throughout the country. Auger moved quickly, and by the end of 1927 had established 823 YPF subagencies with 736 gasoline pumps. Consistent with his interest in promoting Argentine industrialization, Mosconi arranged to purchase these pumps, as well as a variety of other equipment, from Argentina's fledgling national machinery industry. Throughout Argentina, Auger's gas pumps, emblazoned with the national colors and the YPF seal, provided vivid testimony to the fact that within five years of assuming command of the state oil firm the dynamic Director-General had achieved his goal of creating a vertically integrated enterprise.[49]

Despite YPF's impressive growth and expansion, Argentina continued to rely heavily on imported petroleum products during the 1920's, as Table 4.3 shows. In part this was the result of the rapid adoption of motorized transport during the decade, which made consumption of all forms of petroleum products more than double between 1922 and 1928. Though increased domestic production after the opening of the La Plata refinery enabled Argentina to reduce its dependence on imported gasoline and kerosene—by 1930 imports were down to about one-third of consumption—lagging production of crude oil increased Argentina's reliance on imports of crude fuel oil, gas oil, and diesel oil.[50]

The slow growth of Argentine crude output reflected the failure of Mosconi's four-year plan to achieve its ambitious production goals. Table 4.1 shows how far below expectations output at the Comodoro Rivadavia fields fell; by 1927 production had risen only about 45 percent over 1924 levels and was barely a third of what had been stipulated in the plan. The reason for the poor performance was that the original decision to concentrate on intensive drilling on previously explored lands proved counterproductive. What Mosconi had not foreseen was the sharp drop in per-well production coupled with the rise in the water content of what petroleum was pumped out as this period of increased drilling activity wore on (see Table 4.5). Under the four-year plan, exploration of new lands had received only about 4 percent of YPF's investment budget, or enough to sink about six exploratory wells annually; this was the weak link in Mosconi's development plan. When Traian T. Serghiesco, Chief of YPF's Geological Division at Comodoro Rivadavia, charged in 1930 that the enterprise had "lived more for the present than for the future," he echoed criticism that was widespread among Argentine oil experts. Serghiesco revealed that he had strongly opposed the policy of intensive drilling on already developed lands, but that the Director-General had overruled him.[51]

In his official reports, Mosconi overlooked the increasingly disquieting production trends and played down the lack of exploration; instead, he emphasized YPF's diversification, its gains in total output, and especially its healthy profit ratio.[52] Annual profits, which were never below 25 percent and were often closer to 50 percent of receipts during the 1924–27 period, totaled 44.5 million pesos for the four years (see Table 4.6). Mosconi's aim to capitalize YPF as quickly as possible, however, led to several conflicts with the labor force that resulted in walkouts and interruptions of production. Although these disputes and strikes were not as severe as those of 1917–20, their pattern recalled the hostile labor relations that had marred the state oil industry during the first Yrigoyen government.

It cannot be said that markedly higher living standards for oil workers

TABLE 4.4. *State and Private Petroleum Production in Argentina, 1922–1930*
(Cubic meters)

Year	Total domestic production	State fields					Private fields					
		Chubut	Neuquén	Salta	Total	Pct. of domestic production	Chubut	Neuquén	Salta	Mendoza	Total	Pct. of domestic production
1922	455,498	343,910	4,978		348,888	76.6%	106,610				106,610	23.4%
1923	530,209	400,048	7,138		407,186	76.8	122,771	252			123,023	23.2
1924	740,697	544,367	9,554		553,921	74.8	182,211	4,565			186,776	25.2
1925	952,065	609,272	14,764		624,036	65.5	323,872	4,157			328,029	34.5
1926	1,248,032	720,243	23,496		743,739	59.6	492,245	9,825	223	2,000	504,293	40.4
1927	1,371,964	792,364	30,511		822,875	60.0	494,599	51,551	2,861	78	549,089	40.0
1928	1,442,072	811,931	47,532	1,141	860,604	59.7	491,156	73,588	14,424	2,300	581,468	40.3
1929	1,493,067	807,213	61,853	3,105	872,171	58.4	490,295	100,823	29,753	25	620,896	41.6
1930	1,431,107	721,592	89,927	16,494	828,013	57.9	448,127	124,091	30,431	445	603,094	42.1

SOURCE: "Resumen estadístico de la economía argentina," *Revista de Economía Argentina*, 20 (Nov. 1938), p. 323.

TABLE 4.5. *Average Annual Production per Well, YPF Oil Fields, 1923–28*

Year	Wells in production	Crude oil production (m³)	Production per well (m³)
1923	144	407,186	2,827.7
1924	184	553,921	3,010.4
1925	272	624,036	2,294.2
1926	376	743,739	1,978.0
1927	489	822,875	1,682.8
1928[a]	632	859,463	1,359.9

SOURCES: Argentina, [34], pp. 213, 249; "Resumen estadístico de la economía argentina," *Revista de Economía Argentina*, 20 (Nov. 1938), p. 323.
[a] The figures for 1928 do not include the 1,141 cubic meters of oil produced that year in YPF's Salta fields. Compare Table 4.4.

TABLE 4.6. *Financial Results of YPF's Operations, 1923–30*
(Paper pesos)

Year	Sales	Expenses[a]	Profits	Profits as percent of sales
1923	16,662,566	9,662,566	7,000,000	42.0%
1924	18,278,887	10,778,886	7,500,000	41.0
1925	22,476,823	12,476,823	10,000,000	44.5
1926	32,048,466	23,048,466	9,000,000	28.1
1927	50,141,763	32,141,763	18,000,000	35.9
1928	56,686,545	41,686,545	15,000,000	26.5
1929	58,523,437	48,523,437	10,000,000	17.1
1930	65,871,624	55,871,624	10,000,000	15.2

SOURCES: Argentina, [21], pp. 16–20, 36–37; Argentina, [22], p. 26.
[a] In this table "Expenses" includes reserves and amortization; compare Table 2.3.

accompanied the advent of state capitalism in the Argentine petroleum industry. YPF's wages remained low, although they had improved since 1920: for example, in 1924, drilling workers received between 6.08 and 6.72 pesos per day, and production laborers between 5.44 and 5.76. These wages were close to the national average; but the cost of living remained much higher in Patagonia than in the Argentine heartland, so that throughout the 1920's real wages paid to YPF workers remained too low to attract high-quality skilled labor, which was in short supply in Argentina.[53]

Conscious of the dismal conditions petroleum workers faced at Comodoro Rivadavia, Mosconi moved fast to correct the most glaring abuses in housing and working conditions. Immediately after assuming control of

YPF, he replaced the shacks that still housed hundreds of workers with more adequate shelter, and ordered construction of a small modern hospital costing 70,000 pesos. But good medical care still did not receive high priority from the YPF management. As late as 1930, only two doctors staffed the hospital, an inadequate number in view of the high accident rate. Mosconi did alleviate the accident problem somewhat with programs of medical and accident insurance, financed by compulsory worker contributions. He also established paternity and maternity benefits and subsidies.[54]

However, the principal focus of Mosconi's labor relations program was not so much the improvement of living and working conditions as the promotion of Argentine patriotism among the labor force. Convinced that the high number of foreign workers had been the chief source of labor unrest at Comodoro Rivadavia, the Director-General mounted a campaign to recruit and train native-born Argentines, particularly from the poverty-stricken northwestern provinces, to work in the oil fields. But the proportion of native-born workers rose slowly, from about 20 percent in 1922 to 32 percent in 1927, because YPF's rapid expansion meant continued reliance on trained—and hence largely foreign-born—oil workers.[55] In order to instill respect for national values and traditions in the polyglot work force, Mosconi announced a campaign "to implant the principles of Christian morality and the cult of heroes." At Comodoro Rivadavia this program included erection of a stately Catholic church and of a large monument to José de San Martín, which was inaugurated with suitable pomp in 1930.[56] YPF coupled this emphasis on patriotism with strong and unrelenting hostility toward the local labor movement. Colonel Baldrich, Chief Administrator at Comodoro Rivadavia from 1922 to 1924, carried out this antilabor policy with zeal. Baldrich's practice of employing army officers as his chief assistants, and his custom of working in military uniform, symbolized the "militarization" of the labor force to many workers. Moreover, his use of espionage against the labor movement and his prohibition of the sale of working-class newspapers, including the respectably reformist *La Vanguardia*, made him an object of derision to the workers, many of whom believed he knew little about how to run an oil field.[57]

Tension between the administration and the work force reached a boiling point in August 1924, when the Comodoro Rivadavia union, the FOP, called the first major strike since 1920 after the administration responded to FOP demands for improved working conditions by arresting and ship-

ping out two union leaders as "undesirable agitators." About 1,500 workers—two-thirds of the labor force—left their jobs, and oil output ceased.
YPF handled this strike the way Fliess had handled the strikes of 1919 and
1920: the naval transport *América* disembarked a hundred troops, who enforced martial law and ended all organized protest; Mosconi refused to
negotiate and threatened to dismiss all strikers immediately. The result was
that the disgruntled laborers went back to their jobs after a two-week halt
in production.[58] Following the strikes, the YPF administration openly
persecuted the labor movement, prohibited union meetings, and forbade
circulation of union newspapers.[59]

It was three years before the continuing working-class unrest broke out
into another strike. But then in 1927, as part of the Argentine labor movement's determination to protest the execution of Sacco and Vanzetti in the
United States, a general strike was called for August 10 and 11. At Comodoro Rivadavia the plans for a general strike elicited widespread support,
and workers closed down oil operations on August 6. Mosconi reacted
harshly: as the navy's two largest battleships, the *Rivadavia* and the *Moreno*,
stood offshore, marines landed to help the administration break the strike.
The navy severed all communications with the outside world, broke up
union meetings, and imprisoned 200 workers and killed at least one. When
a fire destroyed the local YPF printing plant, Mosconi had over a hundred
workers suspected of anarcho-syndicalist tendencies seized and expelled
from Comodoro Rivadavia.[60] Mosconi broke the strike, but the sullen and
hostile workers began to compare him to the military dictators Carlos
Ibáñez and Miguel Primo de Rivera, and YPF's labor relations remained
one of its weak bases.[61]

The Alvear Government Versus the Oil Companies

While the Alvear government was moving vigorously to back the main
elements of Mosconi's four-year plan, it was also launching a campaign to
restrain the foreign oil companies and to save the country's petroleum
resources for the future use of YPF. The President's aggressive policy, however, reintroduced into Argentine oil politics the delicate and divisive
federalism issue, which as we have seen had prevented Yrigoyen from re-
forming the mining code in 1919. On September 20, 1923, Alvear presented to Congress a comprehensive oil program whose most significant
feature was the proposal to make the national government sole owner of
all mineral rights in the republic. Under the terms of Alvear's program,

control over provincial petroleum concessions would pass to the central government, and YPF, hitherto restricted to the national territories, would be empowered to begin production in the provinces. To soften provincial opposition, the legislation required the national government to distribute royalties to each oil-producing province. Alvear also urged Congress to reform the chaotic system of granting exploration permits to private operators and to empower the Bureau of Mines to levy a tax on private petroleum production.[62]

But the President's oil programs languished in congressional committees. Absorbed in the power struggle between personalists and antipersonalists, Congress spent its time in bitter parliamentary wrangling. After two years passed without action, Alvear resubmitted his proposal and appealed for prompt consideration of it. Denying that his plans implied any sort of "federal monopoly," Alvear pointed out that the private oil companies were rapidly expanding operations in the northwestern provinces but that YPF, under existing legislation, was unable to operate there. But again the legislators did nothing. In November 1925, Alvear determined to convene extraordinary sessions of Congress to deal with the oil question, but these sessions, held during the March 1926 congressional election campaigns, turned into "political and parliamentary chaos."[63]

Meanwhile, convinced that action was essential to prevent the private companies from seizing Argentina's remaining oil lands, Alvear on January 10, 1924, had dramatically exercised his executive powers and issued two sweeping decrees, both based on the 1903 Public Lands Act. The first created a vast state oil reserve totaling about 33 million hectares in the national territories. Only state exploration was to be permitted on unclaimed lands in the new reserve, which included all areas of Patagonia where petroleum deposits were suspected to exist—27.6 million hectares in Neuquén, 4.8 million in Chubut, and 430,000 in Tierra del Fuego. Alvear's second decree ordered the Bureau of Mines to begin a careful, case-by-case examination of pending applications for oil concessions within the new reserve, and then to grant concessions only to those applicants who could prove intent to produce oil. The object was to force the abandonment of claims acquired during the 1920–23 speculative boom by demanding that applicants possess the financial capacity for exploration and by ending the bureaucratic delays that previously had enabled speculators to hold claims for longer than the law technically allowed.[64]

The government moved quickly to enforce these decrees. Responding to

a 1926 congressional inquiry about the status of private oil concessions in
the reserve zone, the President reported that on January 1, 1924, the
number of pending oil concessions in the national territories had totaled
7,236, covering 8.2 million hectares. The Bureau of Mines eliminated
all petitions "whose authors pursue only speculative profit," and by March
10, 1927, had granted only 72 concessions covering about 73,000 hectares.[65]

The government's reserve decrees, along with its strong support of YPF,
touched off a chorus of protest from private oil interests. *Petróleo y Minas*,
the monthly organ of the private oil firms, attacked the decrees not only as
illegal but as contrary to the national interest of increasing oil production
as rapidly as possible. Editorializing that "whoever obstructs production
and promotes imports is a bad patriot," *Petróleo y Minas* also claimed that
the decrees hurt not the "trusts" but the "small operators who abound in
these territories."[66] Nonetheless, the large private oil companies, including Standard, complained that Alvear's policy "almost completely paralyzed any possibility of exploring new zones in this country."[67]

The government's determination to reserve Patagonian oil lands for YPF
also aroused criticism among a number of Argentine petroleum experts,
who believed that increasing production should supersede the question of
national ownership. Perhaps the most influential of these critics was M. J.
Lagos, a retired admiral who published a widely read book on petroleum
policy in 1924. Lagos, who feared that Alvear's policy would lead to a
"state monopoly" under which oil production would stagnate, was convinced that YPF's limited financial resources would prevent it from under-
taking the vast programs of exploration and drilling necessary to increase
production rapidly.[68] His proposed solution was a revival of the mixed-
company concept first suggested as early as 1914; it would resolve the tension between demands for productivity and demands for national ownership
without resorting to a government monopoly. In 1924, however, the top
priority of Alvear and Mosconi was to save at least part of Argentina's
remaining oil resources for YPF's use. Moreover, given the volatile nature
of Argentine politics and the difficulty of institution-building in the oil
sector, the mixed-company plan would have at the least met with serious
delay and probably suffered defeat in Congress during the mid-1920's.
Although Alvear very much wanted to increase production, he acted first
to establish the reserves. Later, in 1927, the President expressed his willingness to support the mixed-company concept, but only on condition that

the national government receive jurisdiction over provincial oil concessions. Mosconi himself later embraced this solution, although he would
tolerate only Argentine capital in the mixed companies (see Chapter 5).

The 1924 decrees touched off a sharp dispute between the government
and the Anglo-Persian Oil Company, which had been operating at Comodoro Rivadavia since 1921. Although the company had invested over £1
million in Argentine operations, by 1925 it had struck oil only twice. After
explorations indicated that rich petroleum deposits existed adjacent to its
holdings, Anglo-Persian requested drilling rights over several thousand
additional hectares. This area, however, fell within the 1924 reserve, and
the Minister of Agriculture rejected the British company's request.[69] In
the same year, 1925, YPF began drilling within 50 meters of Anglo-
Persian's properties. Sir Charles Greenway, Anglo-Persian's Chairman,
journeyed to Buenos Aires and protested strongly to the Ministry of Agriculture that such action would reduce the British firm's output. He complained about the "openly official competition against us," and asked that
YPF cease its drilling. The Minister, however, quickly rejected the British
company's request, pointing out that "the Mining Code sets no limit . . .
as to where the wells are located," and that "if Anglo-Persian requires larger
extensions for exploitation on a vaster scale, the National Operation also
requires the same."[70] Anglo-Persian alienated the Argentine government
by submitting "this case in the form of blunt demands," and received no
British diplomatic support. One does not deal with the Argentine government as though it were "Persian and Eastern potentates," commented
The Review of the River Plate; and British diplomats concurred in this assessment. The Alvear government's decision had the effect of restricting the
giant British oil company to a small role in Argentine petroleum production—a development that the British press criticized but that the Foreign
Office accepted without protest.[71]

The foreign companies believed that Minister of Agriculture Le Bretón
was primarily to blame for the government's aggressive oil policies. Representatives of both Standard and Anglo-Persian complained to U.S. Consul
W. Henry Robertson about Le Bretón's "unfriendly attitude . . . to all
petroleum operations here at present on the part of foreign companies."
According to Bernard S. Van Rensselaer, a U.S. petroleum-industry lawyer, it was clear that Le Bretón, as ambassador to the United States, had
"acquired theories regarding government control and regulation of private

industries which cannot yet be introduced into the Argentine with beneficial results to anyone concerned." [72]

But that the nationalistic oil policy emanated directly from Alvear became clear when Le Bretón resigned in 1925 after a dispute with the President. Alarmed at the disastrous political effects of the schism in the Radical Party, Alvear refused to follow the advice of the militant antipersonalists in his cabinet who urged intervention in the province of Buenos Aires, an Yrigoyen stronghold. Instead, Alvear backed a fusion candidate for the provincial governorship there whom he hoped both Radical factions would support. This policy of conciliation with Yrigoyen caused a cabinet crisis: Interior Minister Vicente Gallo resigned in protest in July 1925, followed by Le Bretón in September. [73] But the foreign oil companies received no respite from the man Alvear named as the new Agriculture Minister—Emilio Mihura, a professional engineer. Mihura, in fact, collaborated closely with Mosconi, who had already launched a major publicity campaign that aimed to stop Standard's rapid acquisition of the best provincial oil lands before Congress could pass national petroleum legislation.

This new attempt to expand the national government's authority over provincial petroleum affairs met the same fate as Alvear's legislative proposals. The national petroleum concession policy that Mosconi and the President believed essential to the future of YPF aroused a basic constitutional conflict that Argentine liberal politics proved unable to resolve. An analysis of the relations among the northwestern provinces, the private oil companies, and YPF may help demonstrate the frustration that surrounded attempts to formulate a nationwide oil policy prior to 1930.

The Arena of Conflict: The Oil Provinces

Oil fever swept the province of Salta soon after the Comodoro Rivadavia discoveries of 1907. Inundated by requests for concessions over the next few years, the provincial government, which allied itself closely with President Roque Sáenz Peña after his election in 1910, suspended all new oil leases in 1911 and decreed a 460,000-hectare petroleum reserve that only the national government's agents might explore. But the provincial oligarchy, dissatisfied with the slow pace of government exploration after 1911, began to urge opening the province to foreign petroleum investment. This prospect quickly gathered political support in Salta, a province whose economy was poor, underdeveloped, and dependent on the sugar industry and on exports of cattle to the Chilean nitrate-mining regions across the Andes. Support for foreign oil investment became particularly strong among the Unión Provincial, the party of the local oligarchy, which regarded petroleum both as a new source of income for the provincial treasury and as a means of curbing Salta's serious unemployment. When Chile's nitrate trade collapsed in 1918 after the German invention of synthetic substitutes, Salta's economy entered a period of deep depression. In the same year, Governor Abraham Cornejo rescinded the 1911 reserve decree and Yrigoyen intervened in the province. However, Yrigoyen's intervenor, rather than tightening restrictions on foreign investment, imposed a set of liberal mineral-land acquisition procedures under which associates of Jersey Standard began to acquire major concessions. In July 1922 Standard organized a large party to explore the oil-rich Orán region, the province's semitropical eastern lowlands, and the following year it leased 240,000 hectares there. By the end of 1924, Standard and its affiliates had acquired control over a total of 1.2 million hectares in Salta province, although no production had yet taken place.[74]

Standard began to encounter political obstacles in Salta in 1922, when Adolfo Güemes, a Radical, was elected governor. Yrigoyen had carried out a second intervention in the province late in 1921, and the intervenor had promoted the candidacy of Güemes, a direct descendant of one of Argentina's great heroes of the Wars of Independence. Güemes, who became a close ally of President Alvear, represented the job-hungry middle class, the core of Radical support in Salta. To gain a popular base, the new governor launched an ambitious public-works program financed by lavish federal aid Alvear pumped into the province. His government's financial stability assured, Güemes began to attack Standard and to argue that the province would receive greater benefits through cooperation with YPF. Soon after taking office, he rejected a Standard proposal to give it exclusive rights over 90,000 hectares of petroleum land for 20 years in return for annual royalties of 9 percent of gross production to the provincial government and one percent to municipalities. After the company attempted to override Güemes by bringing the proposed contract before the provincial legislature, the governor appealed to Alvear for support. On December 12, 1924, Güemes followed the President's example by decreeing formation of an enormous, nine-million-hectare oil reserve that contained all potential petroleum lands in the province. The decree prohibited any new private exploration for five years, but allowed YPF to carry out geological studies and exploration. A companion decree issued the same day stipulated that pending requests for oil concessions would be handled strictly in accordance with the national mining code rather than with the loose procedures in use since 1918. Standard's response was to suspend most of its operations in Salta.[75]

The neighboring province of Jujuy, whose socioeconomic structure closely resembled that of Salta, was another site of intensive activity on the part of Jersey Standard. Working through an affiliate, the firm of Leach Brothers, Standard in 1923 acquired concessions for about 20,000 hectares of potential oil lands in the El Quemado region, near the provincial capital, and began to explore the area. Standard's activities in Jujuy did not encounter serious political difficulties at first, but on December 10, 1924, Benjamín Villafañe, an antipersonalist who had taken office as provincial governor the previous April, issued a decree making an oil reserve of about half the province, including all its suspected petroleum lands *except* the existing Leach concession. During a two-year period, only the national government was authorized to explore the reserve zone; and, like Governor Güemes, Villafañe also decreed tougher procedures for evaluating existing concession requests. Meanwhile, YPF gained a toehold in Jujuy on January 1, 1925, when the State Railways transferred to YPF a 2,000-hectare property near El Quemado on which oil had been discovered in 1923.[76]

Governors Güemes and Villafañe carried out their oil reserve policies in the context of strong pressure from the Alvear government. In 1923, the President, determined to oppose Standard's expansion in the northwest with every means at his disposal, commissioned Mosconi to study the situation and make recommendations. Convinced that Standard held concessions in Salta and Jujuy without having done the necessary work specified by the mining code to retain them, Mosconi sent inspectors early in 1924 to search the provincial mine-office records for evidence of irregularities. The inspectors discovered evidence not only that a number of claims conflicted with the mining code but also that 563 concessions the Salta government had made to various parties since 1918 (totaling 1,126,000 hectares) had fallen into the hands of Standard. Typically, a private party in the employ of Standard would obtain a concession and later cede it to one of the company's subsidiaries or to Standard itself, which thereby gained domain over immense areas of Salta without directly requesting concessions. After Mosconi learned of the extent of Standard's penetration, he urged both Güemes and Villafañe to issue the 1924 reserve decrees.[77]

But in Salta the political scene shifted rapidly and frustrated the plans of Alvear and Mosconi. The 1925 gubernatorial elections took place while the provincial Radical Party was divided between personalists and anti-personalists, and the split enabled the Unión Provincial to return to power under Governor Joaquín Corbalán. A vigorous opponent of what the Salta oligarchy considered the federal government's unconstitutional interference in provincial affairs, Corbalán moved quickly to rescind Güemes's 1924 decrees. Standard then renewed its exploration program and again concentrated its search in the Orán region, where it intensively explored a concession of about 55,000 hectares centered around the town of Tartagal. It brought in its first producing well there in 1926. Reflecting the company's strong concern with its Salta operations, E. J. Sadler, President of Jersey Standard, wrote to the U.S. Secretary of State in November of 1926 that "we feel that this district may well develop into the most important producing area in the Argentine, and that our concessions cover by all means the greatest part of the prospective productive area."[78] By the end of 1928, Standard had invested a total of 15.4 million pesos in the region and had seven wells in production and 13 more in the works. The company built 256 kilometers of access roads, a small hospital for its more than 2,000 workers, and two small refineries. La Nación noted, with some irritation, that only the unskilled workers were Argentine, whereas most of the skilled labor, technicians, and management were North American, Greek, Dutch, and German. Very few of these latter spoke Spanish, and nearly all the company's documents were in English.[79]

Faced with Standard's renewed activity in Salta, the national government mounted a concerted campaign against the province's oil policy. Moreover, Mosconi became convinced that Governor Villafañe in Jujuy was cooperating behind the scenes with Standard by allowing the Leach concession, which contained the richest known oil lands in the province, to remain intact. Alvear pressed Congress to pass his proposed oil legislation in order to bring all provincial petroleum concessions under national jurisdiction, but Congress remained mired in partisan political squabbles and took no action.

Frustrated on the legislative front, the government in 1926 attempted to pressure Corbalán and Villafañe to modify their policies. Alvear punished Corbalán by reducing funds for public works in Salta and by halting construction of a Trans-Andean railway to Chile that was important to the provincial economy. To prod the northern governors further, Mosconi in 1926 lodged a bitter complaint with the Minister of Agriculture charging

14. A conference of interior governors at La Rioja, 1927, which brought together many of the prominent Argentine public figures discussed in this book. Sitting, from left to right, are governors Joaquín Corbalán of Salta, Adolfo Lanus of La Rioja, Benjamín Villafañe of Jujuy, and Agustín Madueño of Catamarca. Among those standing are Luis Colombo, President of the Unión Industrial Argentina (2d from left), and Alejandro Bunge (far right).

that Standard still held its most valuable concessions in both northern provinces and that, in effect, both governors were working hand in glove with the U.S. company. Moreover, he alleged that neither province had enforced the mining code, and requested that the federal government intervene.[80] The Minister of the Interior then made inquiries with Corbalán and Villafañe, but both governors vigorously denied any irregularities. Corbalán defended Standard, claiming that the company's activities were "giving work to thousands of laborers" and were reviving the "anemic economy" of the province from the "deep economic and commercial crisis it had been suffering."[81] Villafañe, well known throughout Argentina for his scathing criticism of what he considered the central government's economic exploitation of the interior provinces, engaged in a long and bitter polemic with various ministers and with Mosconi. Charging that the government's campaign against him was politically motivated, Villafañe emphasized that his administration had eliminated all but 13 of the 400 requests for concessions that had been pending when he had assumed office. The Leach concessions, Villafañe continued, had been made prior to his election and were entirely legal.[82] Mosconi and Alvear were trying to expose and publicize the results that flowed from provincial autonomy over oil concessions. The Leach concession was legal under the terms of the provincial mining code; Mosconi wanted to abolish such provincial rights, and to strengthen his case he portrayed Villafañe as turning over the richest oil lands in Jujuy to the much-despised Standard Oil Company.

Villafañe's conflict with the central government demonstrated the weakness of antipersonalism as a national political movement. Rather than capitalize on provincial resentment against Yrigoyen's export-oriented economic policies, which had helped bring about the Radical Party schism, Alvear also favored the economic interests of the export-oriented regions. In particular, his determined nationalistic petroleum policy divided antipersonalism along regional lines. Faced with the opposition of the President, antipersonalists in the oil-producing provinces began to break with Alvear and to seek allies among conservative groups that also opposed changes in existing oil legislation.

Mosconi, with Alvear's backing, was unwilling to permit Villafañe to defy the national government. The head of YPF charged that the Jujuy governor was not carrying out a demarcation of the Leach concessions, which surrounded YPF's El Quemado holding. Until conflicting jurisdictions were settled, YPF was unable to commence work on its sole Jujuy claim. After the provincial Supreme Court ruled that Standard's concessions were legal, Alvear continued to contest the issue by a decree of November 29, 1926, that ordered YPF to conduct methodical exploration throughout the entire reserve zone of Jujuy.[83]

Mosconi also counterattacked in Salta. He condemned the province's decision to let Standard continue to operate on its Orán concessions, the richest oil lands in the province. Mosconi's ally, the Radical Party newspaper El Intransigente of Salta, joined the campaign against Standard and charged that the company was hiring numerous high provincial officials to represent it. Corbalán answered by reminding Mosconi that the federal government had no constitutional right to interfere in provincial petroleum affairs.[84] Mosconi's determined fight against Standard in the provinces aroused concern not only in the Salta provincial government but also in Standard's New York headquarters, where E. J. Sadler wrote the Secretary of State to express "no little apprehension" over YPF's policy and projected

Argentine oil legislation whose effects would be "confiscatory . . . more through the imposition of impossible conditions than by direct confiscation."[85]

With the conflict over oil in the northern provinces emerging as a major national issue, the personalist Radicals joined Alvear to oppose Standard. Yrigoyen operated from a stronger political base on the petroleum question now than during his first presidency. After the 1924 Radical schism, Yrigoyen no longer faced opposition to a centralizing oil policy within his own party. He quickly abandoned the policy of cautious cooperation with the foreign oil companies and, like Alvear, adopted a hostile attitude toward them and particularly toward Standard. The political position of the oil-producing provinces was becoming steadily weaker, for both Alvear and Yrigoyen supported national jurisdiction over petroleum concessions, though they would differ on the extent of the central government's authority.

By 1926, the personalist Radical Party had adopted a staunchly nationalistic attitude on petroleum affairs. In Congress, personalist deputies warned of Argentina's impending loss of control over oil resources to foreign companies, while *La Epoca* mounted a journalistic campaign against both Standard and the northern governors. Villafañe, long an outspoken critic of Yrigoyen's economic policies, in particular incurred the Radical newspaper's wrath. "There is no criminal irregularity he has not committed, particularly over the question of Jujuy subsoil wealth." Labeling Villafañe "the vulture," *La Epoca* charged that "he has made a calculated delivery of the oil resources to foreign capitalists."[86] Villafañe, no mean opponent in a polemic, answered with attacks on the "Yrigoyenist propaganda campaign," which he claimed aimed "to defame me in the vilest ways." Moreover, the governor began to publish a series of books that exposed corruption within Yrigoyen's government and bore titles such as *Degenerates* and *Yrigoyenism Is Not a Political Party. It Is a National Disease and a Public Danger.*[87]

Rebuffed by the northern governors, the Alvear administration attempted another tactic to give YPF access to petroleum resources in Salta and Jujuy. In 1927, a Spanish citizen, Francisco Tobar, was persuaded to transfer his oil well "República Argentina" and 29 concessions he had held since 1907 to YPF in return for a guarantee of royalties on future oil production. Tobar's properties, all located in Salta's Orán Department, conflicted and overlapped with Standard's claims, which the provincial government never had demarcated precisely. When YPF began drilling on its new acquisition, Standard brought the matter to the attention of the Corbalán government, which ruled in favor of the U.S. company and ordered the state firm to stop.[88] This ruling led to a long court battle beginning in 1928, which the next chapter will examine in detail.

Its campaign against Standard in the northwest stalled, the Alvear administration worked wherever possible to preserve Argentina's petroleum for YPF. In Mendoza province, which contained valuable petroleum resources untapped since the Compañía Mendocina failed in 1897, U.S. interests, probably associated with Standard, acquired in 1927 a Chilean company that held major oil land concessions. This company, El Sosneado, soon thereafter proposed to the provincial government the formation of a mixed company in which each would hold 50 percent of the capital. This proposal, however, foundered in the face of Alvear's determined opposition. The Governor of Mendoza, Alejandro Orfila, was an ally of the powerful Lencinas family, whose reformist and populist policies had made *lencinismo* the strongest party in Mendoza. Originally allied with the Radicals, the Lencinists broke with Yrigoyen in 1922 primarily over the President's centralizing policies. Having incurred the hostility of the personalists, who repeatedly urged Alvear to intervene in the province and destroy the Lencinists, Governor Orfila dared not challenge Alvear when he urged Orfila to abandon the mixed-company plan on the grounds that it imperiled the oil legislation he was urging Congress to pass. As a result, the governor dropped the proposal, and almost no oil flowed from the rich Mendoza fields prior to 1930.[89]

Alvear followed his success in Mendoza by promulgating later in 1927 a series of decrees enlarging the state oil reserve to include nearly all of Patagonia. As in 1924, the President explained that he acted to protect the national interest because Congress had not yet enacted a national oil law.[90]

Alvear's Petroleum Policy: A Summary

During the first five years of his presidency, Alvear carried out a remarkably aggressive oil policy. YPF, little more than a shambles in 1922, became a viable and rapidly growing enterprise. To protect the future of YPF, the government mounted a successful effort to establish oil reserves in the national territories. One major stumbling block to this expansionist oil policy remained—the question of provincial petroleum concessions.

Alarmed that Jersey Standard's activity in Salta threatened to exclude YPF from exploiting that province's rich oil resources, Alvear unsuccessfully introduced national oil legislation and then worked with Mosconi to prod the provincial governments to comply with the administration's policy. In fact, however, Alvear was able to do little more than launch a publicity campaign—which the personalist Radicals joined with gusto—against the recalcitrant northwestern governors.

Alvear's petroleum policy inextricably linked the questions of federalism and provincial rights to Argentine oil politics until political democracy collapsed in the 1930 coup. Convinced that the national government was violating their provinces' constitutional rights, governors Corbalán of Salta and Villafañe of Jujuy defended their autonomous oil concession policies and refused to cancel the claims Jersey Standard already held. The north-western governors identified foreign oil investment with the economic salvation of their poverty-stricken provinces. Like many Argentines from the interior, they distrusted the Buenos Aires bureaucracy, which they believed had long sapped the economic resources of the rest of the country. Villafañe and Corbalán feared that if YPF gained ascendency in provincial production, it too would siphon off wealth to the national government and leave little in return. The lines of political conflict were clearly drawn as Argentina awaited the opening of the 1927 congressional sessions. The northwest governments wanted to control their oil, and this the Radical politicians and YPF's leadership were determined to prevent. For the next three years, this conflict absorbed the nation's attention.

CHAPTER V

Petroleum Politics and the Collapse of Argentine Liberalism, 1927-1930

The year 1930 marks a turning point in Argentine history. Half a century of rapid economic growth, interrupted only by the First World War, came to a sudden halt, and Argentina entered a period of economic stagnation from which it has yet to emerge. In the same year, Argentina's democratic political experiment collapsed into the militarism and authoritarianism that have plagued the country to the present. Argentina in 1930 was highly vulnerable to crisis, for despite the lessons of the First World War the Radical governments of Yrigoyen and Alvear had not successfully confronted the country's most serious problem, its neocolonial economic structure. Tied to a largely urban electorate dependent on the export economy, the Radical presidents in fact had not enjoyed a great deal of latitude on economic policy issues. Only in the petroleum sector did they show themselves willing and able to initiate policies that departed from their general commitment to economic liberalism.

The drift of Argentine economic policy and its subordination to political expediency became particularly clear during the second government of Hipólito Yrigoyen. Elected president in 1928 with one of the largest majorities in Argentine history, Yrigoyen devoted his attention primarily to politics—particularly to enlarging the bureaucracy, rewarding his followers, and strengthening his political machine. Consequently, Yrigoyen regarded the issue of petroleum development less as an economic than as a political matter. Rather than fashioning an oil policy to meet the long-range energy requirements of the Argentine economy, the President strove to nationalize the private companies and to create a state oil monopoly that would provide an attractive source of government patronage. Yrigoyen's decision to make a short-term political response to a fundamental economic question thrust petroleum nationalism to the forefront of Argentine politics. General Mosconi, who remained head of YPF under Yrigoyen, formulated a comprehensive ideology of nationalistic petroleum development that incorporated the mixed-company concept and aimed for both maximum efficiency and national ownership. The Radical chief, however, disregarded Mosconi's concern for productivity and pressed ahead to form a totally state-owned oil monopoly. In pursuing this course, Yrigoyen relied on populist techniques to mobilize support behind his statist version of petroleum nationalism. The result, however, was a disastrous polarization of Argentine politics on the central constitutional issues of federalism and provincial rights. This chapter analyzes the politics and economics of Argentine oil during the late 1920's and focuses on the relationship between the political polarization that accompanied Yrigoyen's petroleum policy and the demise of Argentine liberal politics in 1930.

Argentina and the United States: Economic Tensions

Yrigoyen's national monopoly plan threatened all private oil companies operating in Argentina, but the President's spokesmen directed their criticism against Jersey Standard's Argentine subsidiary and seldom mentioned other companies. The choice of Standard as target does not necessarily mean that the Radicals intended to employ the nationalistic legislation selectively—though the vague wording of the proposed law would have enabled them to do so, and though a 1930 incident involving Royal Dutch Shell suggests that Yrigoyen may have intended to oust only the American oil companies. Whatever the government's intentions, the choice of Standard Oil as the prime object of its attack reflected the bias against both the company and U.S. capitalism in general in Argentina toward the end of the 1920's. Moreover, Standard's cooperation with the Salta and Jujuy governments in their conflict with the national government over oil policy was becoming a major public issue. The company's expansion in the northern provinces only strengthened the Argentine belief—widely held since the denunciations of Luis Huergo and the congressional antitrust investigations of the First World War—that Standard Oil was an unsavory and exploitative "trust." The company also served as a convenient symbol of

American economic penetration, which many citizens, both inside and outside the Radical Party, regarded as a dangerous threat to the national
interest.

Resentment against the United States and its economic policies had deep
roots in Argentine history. During the mid-1920's, though, resentment
grew particularly rapidly because of American protectionism. Throughout
the early 1920's, U.S. exports to and investments in Argentina expanded
tremendously, yet the United States bought relatively little in return. U.S.
exporters relegated Great Britain to second place in the Argentine market
by selling technically advanced products—including office machinery,
mechanical equipment, and especially motor vehicles—at prices the British
could not match.[1] U.S. investments in Argentina, negligible before the
First World War, spurted from 85 million gold pesos in 1917 to 807 million in 1931 while British investments remained essentially static. Although part of this investment financed branch plants to manufacture consumption goods, the bulk of it represented the purchase from British
owners of enterprises that were large consumers of imported equipment,
including electric power stations, tramways, and telephone systems.[2] But
the United States, which was the leading source of Argentina's imports,
ranked only fourth or fifth as a buyer of her exports. Indeed, in the 1925–
29 period, Great Britain bought more than three times the value of Argentine exports than the United States did.

The cause of this unbalanced trade situation was the fact that the Argentine and American economies produced many of the same staples, and that
Argentine linseed, wool, corn, hides, and beef, which could have been
competitive on the American market, were put at a disadvantage by the
high postwar tariffs the United States erected against agricultural imports.
As late as 1913, the United States had admitted eight of Argentina's
principal exports duty free; but by 1922, only hides continued to enter
free.[3]

To correct this serious trade imbalance between the two countries,
Argentines tried to increase the sales of their country's leading export,
dressed beef, in the United States. The product initially proved highly
competitive: in 1924, a San Francisco firm placed an order for 200,000
pounds of Argentine beef and was able to sell it—after paying customs
duties, transport costs, and all other expenses—for ten to fourteen cents a
pound less than meat imported from other countries. Sales of this magnitude gave rise to optimism in Argentina that, as *The Review of the River Plate*

put it, "sooner or later, the United States will become a very heavy importer of Argentine meat." As U.S. Consul Thaw informed the State Department in 1925, "there is quite a prevalent notion that the U.S. can and should increase its consumption of Argentine products."[4] Luis Duhau, president of the Sociedad Rural, the principal cattlemen's association, calculated that if Argentine dressed beef supplied 5 percent of U.S. consumption, Argentine exports of beef would increase by 28 percent.[5]

But Argentina's hopes to increase its exports to the United States were dashed in 1926 when the Coolidge government prohibited imports of chilled or frozen meat from countries where foot-and-mouth disease existed. As a result, U.S. beef imports from Argentina declined from 1.5 million pounds in 1926 to nothing in 1927. Argentines responded with a storm of outrage at what they considered an affront to the quality of their principal export. Government officials, cattle exporters, and the press protested that the American restrictions were unfair and politically motivated, and to substantiate their case they pointed to Great Britain's decision to continue Argentine meat imports under more careful inspection and certification procedures.[6] "One cannot accept the North American decision as a friendly act," editorialized *La Prensa*, which viewed the restriction as "a threatening sword against our cattle industry" based on "doubtful scientific fact."[7]

The U.S. restriction, along with tariff increases among such traditional Argentine customers as Spain, France, and Italy, left Great Britain more important than ever to Argentine exporters. But Argentina's traditional free access to the British market no longer was secure. Between 1925 and 1929, as W. Arthur Lewis has written, "there was not even an interlude of prosperity" to break the gloom of serious unemployment, excess industrial capacity, and general economic malaise that shrouded the British Isles. The depressed British economy gave rise to the powerful Imperial Preference Movement, which aimed to impose tariffs on goods imported from outside the Empire. Aware that no country would suffer more from Imperial Preference than Argentina, the British government began to use the threat of protectionism as a lever to recapture its lost position as the leading supplier of Argentine imports, and thus to redress the balance of Anglo-Argentine trade, which had begun to run heavily against Great Britain.[8] The British ambassador in Buenos Aires, Sir Malcolm Robertson, assured the Foreign Office in 1927 that "this market can help considerably towards relieving unemployment in England, the whole object and end of all that this

Embassy is endeavouring to do in this country." In the same year he publicly suggested that Argentina's best interests favored a trade policy of "Buy From Those Who Buy From Us."[9] The Sociedad Rural quickly adopted this slogan and launched a political campaign designed to protect the republic's position in the British market in return for lower tariffs on imports from Britain. While Luis Duhau emphasized that American trade policy left Argentina no choice but to seek a closer relationship with the United Kingdom, the Sociedad's publications were urging the government to shut off imports from the United States as quickly as possible.[10]

As the dispute over trade deepened, hostility and resentment against the United States swept Argentina. By 1927, the nation's most respected newspapers not only were condemning American economic policy but were adopting a critical attitude toward the United States in general. American movies shown in Argentina, for example, were labeled "decadent" purveyors of an alien system of values. After Washington sent troops into Nicaragua in 1927, U.S. Latin American policy received scathing condemnation both in the press and in Congress. The Nicaraguan intervention, warned critics in *La Prensa* and *La Nacion*, demonstrated that American trade and investment were the harbingers of outright military occupation.[11] The execution of Sacco and Vanzetti further tarnished the image of the United States, particularly among the labor movement. As we have seen, to protest the fate of the two anarchists, the unions mounted general strikes that halted Argentine commerce and transport on three separate occasions in August 1927.[12] Late in 1927 and early in 1928, the mood of anti-Americanism took a violent turn when bombs, one of which injured 21 people, exploded in two U.S. banks and at the U.S. embassy.[13] From a political viewpoint, then, anti-American feeling in 1927 provided a favorable atmosphere for an attack on Jersey Standard.

The State Oil Monopoly Plan

Although the Chamber of Deputies had failed repeatedly to consider Alvear's proposed petroleum legislation, it stirred to action during the 1927 sessions. This unaccustomed activity on the part of the legislators followed Yrigoyen's decision to run for the presidency in the 1928 elections. Now 76, the Radical leader had maintained his popularity and his leadership among the party faithful by continued close attention to organizational detail and by implicit promises of a return to the plentiful patronage policy that had characterized his first administration. YPF's well-publicized campaign against Jersey Standard and the northwestern governors gave Yrigoyen the issue he needed to return to the Casa Rosada. But the Radical chief opposed Alvear's oil proposals, which centered on the idea of extending the federal government's jurisdiction over mining concessions to the provinces. Instead, Yrigoyen instructed his followers in Congress to formulate a more advanced plan that would nationalize the petroleum industry and place it under a YPF monopoly. This oil-monopoly strategy gave the personalists "an ideal popular slogan," as David Rock has put it. Since Alvear's federal-jurisdiction plan enjoyed support among Socialists and antipersonalists, adoption of the state-monopoly idea was politically advantageous for the personalists, who rallied behind Yrigoyen's portrayal of himself as the leader of a nationalistic economic policy that promised to benefit much of the electorate.[14] This new oil proposal quickly aroused attention throughout the nation. As the newspaper *El Intransigente* of Salta correctly noted, "Oil is the topic of the day—no more interesting affair has arisen in the last twenty-five years."[15]

Argentine petroleum affairs in the late 1920's provide an early Latin American case study of a party's use of a nationalist economic issue to gain and consolidate political power. Yrigoyen set out to generate support for his party by promoting the statist version of petroleum nationalism among the middle classes, which by this time were at least a third of the country's population. By 1927, Argentina was predominantly an urbanized and literate country with a rapidly growing educational system. The expansion of the educated population, however, was exceeding the capacity of the export-dependent economy to create suitable middle-class jobs. The resulting white-collar underemployment—a nagging problem in twentieth-century Argentina whose severity has increased over the years—created a large pool of frustrated and resentful people to whom Hipólito Yrigoyen appealed.[16] Yrigoyen did not direct his petroleum campaign primarily to the working class or the labor unions, for the predominantly foreign-born workers of Buenos Aires and the other large cities had not taken Argentine citizenship and remained without the vote.* Instead, he formulated his oil policy with an eye toward the growing Argentine-born middle classes

---

*In fact, the anarchosyndicalist and socialist trade unions, including YPF's own FOP, remained openly hostile to a state monopoly over the oil industry. Not until the presidency of Juan Perón did YPF's labor force support petroleum nationalism.

in the large cities where petroleum consumers were concentrated and where the promise of increased government employment was a potent political appeal.

Yrigoyen's petroleum campaign supports Harry G. Johnson's theoretical conception of economic nationalism in a developing country as a political means to redistribute income by creating new jobs for a marginalized middle class. Despite the fact that oil is not a labor-intensive industry, Yrigoyen's nationalist program caught the imagination of the Argentine middle sectors. There is no evidence that Yrigoyen planned to extend nationalism to other areas of the economy, but the anti-imperialist and populist rhetoric the Radicals employed may have led Argentines to interpret the petroleum campaign as only the first step in a movement to transfer control of Argentina's wealth from foreigners to the middle class by means of state intervention. This is precisely the charge that Conservative opponents of Yrigoyen's oil policy laid against the President.[17]

The strong support of Argentina's student population for Yrigoyen's statist petroleum plan exemplifies the appeal of petroleum nationalism among the middle classes. As in Rumania during the same period, where Maurice Pearton has pointed out that "the frustrated aspirations of the Rumanian technical intelligentsia were one facet of the resurgence of nationalism in Rumanian oil affairs," in Argentina the predominantly middle-class students identified economic diversification and development with the expansion of employment opportunities and the advancement of their careers. The strong anti-imperialist emphasis of the Argentine student movement throughout the 1920's reflected the anxiety about career frustration that students felt within the dependent and nonindustrialized economic structure. Thus when the oil issue surfaced, the Federación Universitaria, the principal student organization at the University of Buenos Aires, became a strong and vocal supporter of Yrigoyen's petroleum policy. Throughout the capital, the Federación organized public meetings, distributed leaflets, and lobbied vigorously to alert the public "to protect our rich subsoil resources from the avarice of capital." Studies at technical and industrial schools, which had begun to train petroleum technicians in 1926, joined the university students in directing a steady stream of petitions to Congress in support of a state oil monopoly.[18]

A pioneer in techniques of popular mobilization, Yrigoyen directed an emotionally charged campaign to rally the electorate behind the petroleum monopoly plan.[19] Radical Party newspapers constantly devoted front-page

15. A group photograph featuring General Alonso Baldrich, seated 2d from right. Leopoldo Lugones is sitting 2d from left. The exact date of this photograph is not known, nor is it known where it was taken.

attention to the oil question between July and September of 1927. Headline after headline in *La Epoca* denounced "the invasion of imperialist capitalism," warned that Alvear's proposal was "defrauding national aspirations," and assured the public that the Radical Party aimed "to defend the public wealth."[20] "Choose, compatriots!," exclaimed one writer in *La Epoca*: "Either a sea of oil in which our sovereignty is drowned, or the sacred sovereignty afloat, if only in a pond of petroleum!" The same newspaper carefully publicized street meetings at which personalist leaders thundered denunciations of Standard Oil and defended Yrigoyen's nationalization plan as the purest embodiment of patriotism.[21]

No supporter of the Yrigoyen petroleum plan was more vocal than General Baldrich, who became one of the leading advocates of a state petroleum monopoly. Significantly, Baldrich served as head of the Army Engineer Corps, a group active in the development and management of YPF. The outspoken general was an effective public speaker who lectured in theaters, delivered addresses over the radio, and appeared before military groups such as the Naval Officers' Club tirelessly repeating that the foreign oil companies endangered the Argentine economy and were "mortifying to the national honor."[22] Baldrich attracted wide attention by publicizing a 1926 incident in Salta in which Standard Oil allegedly attempted to take justice into its own hands. On May 25, 1926, two Standard engineers had been robbed and killed in a remote region of Orán Department. Posters soon appeared throughout the region announcing Standard's offer to pay a reward of 5,000 pesos for each of the assassins, dead or alive. Governor Corbalán's administration hastened to arrest six Argentines, one of whom died in jail and all of whom, Baldrich claimed, were tortured. The company later apologized to Corbalán and alleged that the posters were not officially authorized, but Baldrich seized on this incident to fortify his point that Standard disregarded Argentine law and constituted a dangerous provocative element in Argentine politics. *La Prensa* agreed and published strong editorial condemnation of Standard's conduct.[23] The affair's conclusion proved particularly embarrassing for Standard: after 26 months in jail, the five Argentine prisoners were released when the Salta government revealed that two American citizens were the culprits.[24]

Faced with this barrage of criticism, Standard attempted to defend its position in a long statement to the press in August 1927. Defending itself as a good corporate citizen, the company pointed out that since its incorporation in 1922 it had invested about 25 million pesos in Argentina, at least six million of which had gone for wages and salaries. The present payroll of over 2,000 included only 85 U.S. citizens; the "immense majority" were Argentines. Standard claimed to have helped develop the provincial economy of Salta not only by putting men to work but also by building roads and hospitals in remote regions. The company vigorously denied the claims of its critics that it did not intend to produce oil in Argentina.[25]

The other private oil companies aligned themselves solidly with Standard to oppose changes in the government's petroleum policy. In a petition to the Chamber of Deputies, the major Argentine and foreign companies emphasized that the national interest required increasing production as rapidly as possible and that restrictions on private investment would conflict with that goal. The companies pointed to YPF's lagging output as evidence that the state firm was technically, administratively, and financially incapable of running the oil industry effectively. In a less restrained tone, the journal *Petróleo y Minas* dismissed the monopoly plans as "lunatic fantasies."[26]

The petroleum companies' arguments coincided with the interests of the oil provinces' political elites, who remained unalterably opposed to changes in Argentina's petroleum legislation. Although support existed in the provinces for the state monopoly plan, primarily among the Yrigoyen faithful, the provincial elites raised the banners of federalism and defense of provincial constitutional rights. The theme that the petroleum monopoly would accentuate the central government's exploitation of the interior also appeared frequently in provincial protests against the proposed oil legislation. Writing in the respected *Revista Argentina de Ciencias Políticas*, Atilio Cornejo of Salta complained bitterly that the Radical Party's oil policies not only trampled on the provinces' constitutional autonomy but also represented another step in the long process through which Buenos Aires had absorbed the interior's riches and had given only "alms" in return. Petroleum nationalization, wrote Benjamín Villafañe, would fill the federal bureaucracy "with professionals from the basest political groups" and would deal Jujuy a "death blow" that would "condemn it to eternal misery and depopulation." José Hipólito Lencinas of Mendoza similarly charged that the *porteño* elites thought of little else but enriching themselves with the provinces' wealth.[27]

Although centered in the interior, opposition to the petroleum legislation also emerged among Conservatives in the littoral provinces. For example, Matías G. Sánchez Sorondo, an aristocratic Conservative political leader who was also Professor of Mining Legislation at the University of La Plata and legal adviser to Standard Oil, published in 1927 a book attacking Yrigoyen's oil legislation entitled *La palabra de un patriota sobre el problema de la legislación del petróleo*. An articulate exercise in constitutional law, this book argued that although the state had the constitutional right of "vigilance and regulation" over mines, the constitution clearly prohibited the state from actually declaring them its property.[28] Sánchez Sorondo's concern sprang from his conviction that oil nationalization represented the first step in a movement to abrogate the right of private property. Denouncing the oil law as "ill-advised, revolutionary, and anarchical," he warned that "yesterday it was rents, today it is oil, tomorrow it will be rural lands threatened by redistribution. At base, this is a war against the social structure." Opposed to the federalization plan as well as to the monopoly one, Sánchez Sorondo warned that the "monstrous" oil legislation "would force the provinces to disappear as political and territorial entities."[29] Sánchez Sorondo's book should have been a clear warning to

Yrigoyen that Conservatives would view any government that tampered with private property as illegitimate.

Although debate polarized around the monopoly plan, a few prominent voices, including *La Prensa*, urged Congress to compromise and to adopt a policy combining federal jurisdiction with mixed private-state companies. The most important group to support this position was the Unión Industrial Argentina (UIA). Historically opposed to Argentina's free-trade policy and a proponent of protective tariffs and other measures to promote industry and economic self-sufficiency, the UIA viewed the oil question as a matter of the greatest concern. The industrialists had long hoped to participate in the development of Argentina's oil resources, and since 1914 had lobbied for a mixed-company plan on the grounds that it would best combine the goals of maximum productivity and national control. At a public lecture in 1927 before an audience of 1,200 people, including the ministers of War and Agriculture, UIA President Luis Colombo emphasized that integration of the national capitalist class into the state oil company would promote Argentina's industrial future as well as its "economic and political independence."[30] Alvear himself, as we shall see below, came to back the mixed-company concept in 1927; nontheless, this "halfway" formula, however defensible in logic, aroused only limited political support when confronted by Yrigoyen's plan for a state oil monopoly. The determination of the personalist Radicals to return to power on the oil issue made compromise on the mixed-company plan politically impossible.

Whereas industrialists solidly opposed the idea of a state monopoly, the Socialists divided over the issue. During the 1927 congressional sessions the party suffered a schism—resulting primarily from personal disputes among the leadership—that led to the formation of two Socialist parties, the second calling itself the Partido Socialista Independiente (PSI). The new party, though it controlled 11 of the 19 Socialist seats in Congress, needed an issue to distinguish itself from its rival. Since it hoped to build an electoral base in the capital, where sentiment favored the oil monopoly, it consequently adopted a position in August 1927 supporting the monopoly plan.[31] Meanwhile, the remnants of the original Socialist Party, which had always viewed Yrigoyen as an enemy, remained unalterably opposed to the personalist Radical petroleum policy. The Socialist organ *La Vanguardia* charged that Yrigoyen's oil policies were motivated purely by electoral considerations, that the Senate never would approve the legislation in its present form, and that the net result of the Radical policy would be to

16. "Creole Derivatives of the Petroleum Monopoly," cartoon from *La Van-guardia*, September 12, 1927. The Socialists used "Creoles" as a disparaging term to refer to Yrigoyen's followers, who allegedly employed traditional, "barbaric" political methods. The Socialists wanted Argentine politics to follow "civilized" European models and to abandon these "Creole" traditions. The papers flying out of the spigot are the derivatives, and include "bureaucracy," "votes," "deficit," and "appointments." The two groups tied up are "Yrigoyenists" and "Independent Socialists"; apparently we are to infer that Yrigoyen (in foreground) has successfully tied them to his "oil policy machine."

leave Argentina without any oil law at all. A state oil monopoly would produce "financial disaster, economic ruin, and a formidable instrument of political and electoral corruption."[32]

Petroleum Politics and Yrigoyen's Return to Power

While Argentina's interest groups and political parties were defining their positions on the petroleum question, the Chamber of Deputies prepared to begin debate. Each attempting to gain maximum political advantage, the personalist Radicals and the opposition parties worked through two different committees to prepare alternative programs of oil legislation for consideration on the floor of the Chamber. The two committees involved,

Commerce and Industry on the one hand and General Legislation on the other, had been investigating the petroleum question since 1923 and had developed a string of legislative proposals, all of which had died in the atmosphere of partisan bickering that wracked Congress during the Alvear presidency. During the 1927 sessions, the first package of petroleum legislation presented to the Chamber was prepared by the Committee on Industry and Commerce. This proposal, Chamber Dispatch No. 95, reached the floor on June 13. Basically a restatement of Alvear's earlier plans, it enjoyed support among antipersonalist and some Socialist deputies (this was before the Socialist schism) as well as among some Conservatives from the littoral provinces. Dispatch No. 95 proposed to give jurisdiction over provincial oil, coal, and iron concessions to the national government; to maintain the right of private investors to participate in the oil industry; to authorize the President to form mixed private-state oil companies; to create a state monopoly over the land transport of oil; and to obligate all private operators to pay production taxes, part of which the federal government would return to the provinces.[33] Before the Chamber began debate, however, the General Legislation Committee issued a draft of another petroleum proposal, Chamber Dispatch No. 77. This draft would have established not only federal jurisdiction over mineral deposits throughout the republic but also the principle of state ownership of such deposits, including coal, iron, and oil. Shortly after the introduction of Dispatch No. 77, the personalist bloc in the Chamber received "direct and precise instructions from Yrigoyen" to support it but to amend it with provisions establishing a national oil monopoly.[34]

The Chamber's petroleum debates began on July 28 with a bitter dispute over which of the two committee reports should have precedence. Deputy Diego Luis Molinari, leader of the personalists in the Chamber, began with a vehement attack on foreign oil companies and a demand that Dispatch No. 77 be considered first. A man of combative spirit, an excellent public speaker, and one of Yrigoyen's most trusted subordinates, Molinari consistently supported the state monopoly plan. His populist rhetoric and the key role he played in the petroleum debates, both in 1927 and later, led some observers to believe that he aspired to the presidency.[35] But on this occasion his rhetoric failed to move the Chamber, where the personalists held only 59 of 158 seats. (The antipersonalists held 29 seats, the Conservatives 44, the Socialists 19, and minor parties 7.) A coalition united in opposition to the personalists voted to debate Dispatch No. 95. Molinari and his followers walked out of the Chamber in protest, but since a quorum still existed and the debate continued, the personalist deputies returned a few days later.[36]

Lengthy and often irrelevant speeches punctuated by complicated parliamentary maneuvers characterized the debates, which continued through late September. Dozens of deputies analyzed every conceivable aspect of the oil question and gave long dissertations on political and economic theory citing such varied "authorities" as Montesquieu, Rousseau, Mill, Kautsky, Lenin, Mussolini, Borah, and Coolidge. Antipersonalist and Socialist deputies spoke strongly in favor of the federalization plan and based their argument on the premise that the 1853 Constitution gave the central government power to draft a mining code. Minister of Agriculture Emilio Mihura attempted to placate provincial opposition to Dispatch No. 95 with assurances that the federal government would return oil royalties to the provinces.[37]

The personalist strategy during the debates was to support the idea of federal jurisdiction but to attack the remainder of Dispatch No. 95 on the grounds that the mixed-company provision would allow foreign companies to control the Argentine petroleum industry.[38] When they took the floor, personalist deputies lauded the state monopoly plan and punctuated their speeches with patriotic rhetoric. Molinari, for example, exclaimed: "The oil that God gave Argentina belongs to the Argentines and is for the Argentines!" He emphasized that the state monopoly would provide greater opportunity for "native-born technicians, native-born engineers . . . who speak our language and not English, and who do not take orders from North America."[39] Socialist Nicolás Repetto replied with accusations that personalists were in fact attempting to capitalize on the prevailing Argentine hostility against the United States for sheer electoral purposes.[40]

With the exception of a few personalists, deputies from the oil-producing provinces condemned the petroleum proposals of both Alvear and Yrigoyen. Claiming that Dispatch No. 95 was unconstitutional, they argued that because the provinces had existed before the establishment of a national government, the provinces held prior claim over disposition of their mineral resources. But aside from Conservatives, who were divided over national jurisdiction, few deputies listened when members from Salta and Mendoza attacked what Lencinas called the "robbery" of their rights and portrayed the sad economic state of the interior, which they charged was the result of the federal bureaucracy's economic exploitation.[41] Finally, on September 1, the Chamber voted to give general approval to Dispatch No. 95. Then, by a vote of 88 to 17, the deputies approved Article 1 of the Dispatch, which established federal jurisdiction over petroleum resources, gave the state a monopoly of oil transportation, and prohibited petroleum exports. Exhausted, the deputies then adjourned for a week's respite before voting on the remaining articles.[42]

When debate resumed, the personalists and their Independent Socialist allies introduced an amendment to Article 1 designed to establish the principle of a national oil monopoly throughout the republic. Although existing private companies would not have faced expropriation under the amendment's terms, YPF alone would have been empowered to carry out all new exploration and production. The opposition parties protested, but the adherence of the 11 PSI deputies gave the personalists a 65–55 victory to approve the amendment on September 8. Analysis of the balloting (Table 5.1) reveals that the vote closely corresponded to party alignments. In regional terms, deputies from the Federal Capital voted heavily in favor of the amendment, deputies from the interior voted just as heavily against it, and the deputies from the littoral provinces were divided. Following the vote of September 8, the Chamber quickly approved the remainder of the oil legislation, which dealt with the authority and organizational scheme of YPF. The private oil companies reacted bitterly to the Chamber's action, which, editorialized *Petróleo y Minas*, "is ruinous tyranny on the Soviet model." Ominously, the petroleum industry journal added that the legislation promised "to provoke the ruin of the oil industry and of the country's civic institutions."[43]

The petroleum debates of 1927 coincided with the opening of the presidential election campaigns for 1928. During his final months in office, Alvear refused to give the lackluster ticket of the antipersonalist Radicals Leopoldo Melo and Vicente Gallo any practical assistance. Although they campaigned vigorously and had the support of the Conservative party, they attracted little interest compared to Yrigoyen, who stayed completely out of the public view. Characteristically, Yrigoyen suggested no policies and made no public speeches. Instead, as the London *Times* later put it, he "moved in mysterious ways, creating behind a veil an aspect of deity." The Radical press skillfully promoted Yrigoyen's image as a quasi-divine friend of the common man during the election campaign. "He is the *caudillo* of freedom," *La Epoca* lyricized in February 1928. "The magic sorcery of his heart conquered the hearts and minds of men as he swept through the

TABLE 5.1. *The Chamber of Deputies Vote to Create a State Petroleum Monopoly, September 8, 1927, by Party and Region*

Party	Federal Capital		Littoral provinces[a]		Interior provinces		Totals	
	Yes	No	Yes	No	Yes	No	Yes	No
Socialist	0	8	0	0	0	0	0	8
Independent Socialist	9	0	0	0	2	0	11	0
Radical (personalist)	14	0	29	3	6	4	49	7
Radical (antipersonalist)	0	0	4	5	0	7	4	12
Progressive Democratic	0	0	0	1	0	4	0	5
Conservative[b]	0	0	1	16	0	7	1	23
TOTALS	23	8	34	25	8	22	65	55

SOURCE: Argentina, [2], Sesiones ordinarias, vol. 4 (Sep. 8, 1927), p. 478.

NOTE: From the total breakdown of seats held by parties in the Chamber of Deputies given on p. 124, one can see that the monopoly plan could have been defeated if more Conservatives and antipersonalist Radicals had been present. Let me make three points by way of a possible explanation for the result: (1) absenteeism on roll-call votes was traditionally high; (2) opponents of the monopoly plan knew they could stop it in the Senate anyway; and (3) some Conservatives and antipersonalists may have favored the idea but may not have wanted to go on record as saying so (and thus agreeing with Yrigoyen).

[a] The "Littoral provinces" include Buenos Aires, Entre Ríos, Santa Fe, and Corrientes.

[b] A number of provincial conservative parties of varying titles are grouped under the rubric "Conservative."

plains."[44] While Yrigoyen avoided the limelight, the personalist machine skillfully used the oil policy debates of the previous year to present the masses with an image of a Radical Party that was leading the fight to prevent the rapacious Yankees from exploiting Argentina's petroleum resources.[45]

The election, held on April 1, 1928, gave Yrigoyen the greatest triumph of his long political career. He won 838,583 popular votes, or 57.4 percent of the total, whereas his closest opponents, the Melo-Gallo ticket, won only 28.3 percent. The Socialist candidates took only 4.4 percent of the vote. Personalist Radicals won a clear majority of 92 seats in the Chamber of Deputies as well in the April election, as Conservatives dropped to 36 seats and the antipersonalists to 15. The Socialists declined from 19 seats to ten, with the PSI outpolling the establishment Socialist Party in the capital to win six of the ten seats. The rest of the Chamber's seats fell to the Progressive Democratic party. Though predominant in the Chamber, the personalist Radicals still held only seven of the Senate's 30 seats, far short of a majority.[46] Analyzing Yrigoyen's stunning electoral victory, the *Buenos Aires Herald* commented that his supporters "are convinced that he will set about, immediately after his ascent to the presidential chair, to curb the rapacity of the bloated foreigners."[47] In reality, however, the new Radical government adopted a stance that differed greatly from the image of economic nationalism the President and his supporters presented to the public. Yrigoyen in his economic policies strove for ever-closer relations with Great Britain and directed his nationalism against only one target—Standard Oil.

Buoyed by their party's victory and their secure control over the Chamber of Deputies, the personalist Radicals continued to emphasize the petroleum issue when the 1928 congressional sessions opened. They proposed new legislation designed to enable the President to expropriate the property of existing private oil companies and to grant YPF a complete monopoly over the Argentine petroleum industry. The Radicals were aware that the Senate would reject this plan, for in the Upper House each province held two seats and the interior thus had far greater weight than it did in the Chamber. Indeed, far from searching for a compromise acceptable to the Upper House, the personalists hoped that the Senate's rejection of Yrigoyen's legislation would discredit its members and open the way to election of Radical senators—through provincial intervention if necessary. Yrigoyen's grand political design—control of both houses of Congress—thus became inextricably linked both to petroleum politics and to the federalism issue.

When the personalists in the Chamber introduced the expropriation legislation in July 1928, they touched off a vigorous protest. *La Prensa*, which had supported the concept of federal jurisdiction in 1927, editorialized that "we believe it impossible to consider the projected law seriously," and asked for study and clarification of such difficult questions as how much the expropriations would cost and who would finance them. At the other end of the political spectrum, *La Protesta*, voice of the anarchist trade unions, reviewed YPF's labor relations and concluded that "the State is the worst boss and a very bad administrator." Alejandro E. Bunge, whose reputation as a critic of economic dependence was by this time well established, was appalled at the proposal and published an article charging that expropriation was unnecessary, would force oil production to decline, and would actually impede industrial growth rather than promote it. Again, as in 1927, provincial leaders raised their voices to protest what Lencinas called "the centralization of wealth in Buenos Aires from this immense resource."[48]

As in 1927, the personalist Radicals attempted to mobilize public opinion behind the expropriation proposal, although Radical street orators and student organizers did not need to prod the Chamber to act.[49] The short debate over the expropriation proposal demonstrated that Yrigoyen's oil policy was uniting the opposition parties into a solid front: even the Independent Socialists, who had supported the monopoly principle in 1927, abandoned the Radicals and charged that the expropriation legislation was financially irresponsible. PSI leader Antonio de Tomaso estimated that Yrigoyen's plan would obligate the government to pay 120 million pesos in compensation, and he argued that this vast sum would be better spent to develop YPF's present holdings. But the Radical majority brushed off such mundane objections and on September 17 approved the legislation by a vote of 79–17. Demoralized by the Radical majority in the Chamber, few opposition deputies bothered to appear for the balloting.[50]

Only the Senate remained as an obstacle to Yrigoyen's oil policy, and during its 1928 sessions the Upper House demonstrated strong opposition to the monopoly and expropriation plans. The personalist leader Molinari, elected a senator from the Federal Capital in 1928, delivered a stirring anti-imperialist address warning that Standard Oil's investments presaged the American economic conquest of Argentina. But Molinari's colleagues refused his request to place petroleum legislation on the agenda and instead voted to postpone consideration pending further study of the issue. *La Epoca* viewed this decision as intolerable obstructionism and exclaimed that the Senate would have "no excuse at the bar of public opinion should it persist in this passive attitude."[51] The lines of political conflict now were clearly drawn: in order to obtain a state oil monopoly, Yrigoyen would have to gain control of the Senate; and in order to change the composition of the Senate in his favor, Yrigoyen would have to pursue a course that threatened to produce a fundamental political crisis and the collapse of the government's legitimacy.

Mosconi and the Ideology of Petroleum Nationalism

While the politicians were reaching an impasse over petroleum legislation, the most influential oilman in Argentina, General Mosconi, was working to arrange a compromise solution. Although Mosconi had remained silent during the 1927 petroleum debates, by 1928 his apprehension over Yrigoyen's oil policy became clear and was mentioned by the opposition during the expropriation debates.[52] The Director-General of YPF had by no means abandoned his hostility toward the foreign oil companies; but as a pragmatist he regarded the President's expropriation and state-monopoly plans as ill-conceived. Indeed, Mosconi had dramatically publicized his continuing commitment to nationalistic oil policies during a major inter-American trip he undertook late in 1927 and early in 1928. Consistently, the scrappy

17. General Mosconi, 1928.

general condemned the petroleum multinationals, portrayed YPF as a model for other oil-producing countries, and urged the Latin American nations to formulate cooperative oil-development policies that would reduce the region's dependence on foreign petroleum investment.

Mosconi left Buenos Aires on December 15, 1927, and began with a short visit to the United States to procure machinery and equipment for YPF. The chill that pervaded U.S.-Argentine relations was much in evidence during Mosconi's stay in New York. The Coolidge administration officially ignored his presence, and Mosconi in turn told reporters that he would not purchase oil equipment in the United States were it available elsewhere, that the Nicaraguan intervention was an outrage, and that Washington's economic policy in Latin America would have unfortunate long-term results.[53] In sharp contrast to the cool reception the general received in New York was the fervent welcome the Mexican government

18. General Mosconi delivering an address at the Colegio Militar, Mexico City, 1928.

prepared for him. Arriving in Mexico City in the midst of a bitter dispute between Mexico and the United States over petroleum legislation the former had recently approved, Mosconi conferred with President Calles and delivered major policy addresses at the National University and the Military Academy. He lashed out at the international oil companies, explained the progress of YPF, and urged the Latin American governments to cooperate to fight the "exploitative trusts." The Mexican visit was a great success and marked the high point of Mosconi's tour. The rector of the University entertained him at a splendid banquet, the press lauded him, and Mexican nationalists carefully noted his message. The Mexican visit reverberated both in Washington and in Buenos Aires, where U.S. Ambassador Robert W. Bliss called on the Argentine Foreign Ministry to present a protest against the criticism of U.S. Latin American policy made by the head of YPF.[54]

Another warm welcome awaited Mosconi at his next stop, Colombia.

Invited as the guest of the government, Mosconi arrived in Bogotá while the National Congress was engaged in heated debate over tough new oil legislation proposed by the anti-American Minister of Industry, José A. Montalvo, and aimed largely against a controversial concession a Gulf Oil subsidiary had gained in 1926. As in Mexico, the Argentine general met with high government officials, including Montalvo and President Miguel Abadía Méndez. At a state banquet hosted by Montalvo, Mosconi again recounted the history of YPF and urged a united Latin American effort against the international oil companies. As Mosconi later explained, his purpose was "to suggest to the governments of all the Latin American countries the enormous benefits that would result if each country followed the lines of the Argentine experience."[55] But the head of YPF received a cool welcome at Lima, his fourth stop. Peruvian President Augusto B. Leguía, who opposed state-owned oil companies and who a few years earlier, in a decision much criticized in Peru, had delivered the rich La Brea–Pariñas oil fields to an affiliate of Jersey Standard, granted Mosconi an interview and heard him out but dismissed him without comment. On his final stop, in Chile, the Argentine general received a more cordial response from President Carlos Ibáñez, also a career army officer. Although Chile as yet produced no petroleum, intense foreign interest in exploring the far-south Magallanes region had prompted the Ibáñez government to enact a series of laws that effectively created a state monopoly over oil production. Encouraged by Ibáñez's interest in YPF, Mosconi departed for Buenos Aires in mid-March 1928.[56]

After his return to Argentina, Mosconi not only resumed his role as chief administrator of YPF but also devoted his attention to formulating and expounding a realistic ideology of petroleum nationalism. He drew on his experience as head of YPF to develop a program of petroleum development that he believed to be both politically feasible and consistent with the goals of industrialization and economic self-sufficiency for Argentina. In doing so, Mosconi became the first Argentine or Latin American to publicize an integrated set of assumptions, theories, and goals regarding state owner-ship and control of petroleum resources. As a result, his ideas had (and continue to have) a powerful impact on government oil policy, both in his homeland and in much of the rest of Latin America.

Mosconi founded his oil program on the firm belief that the goals of industrialization and economic self-sufficiency were vital to Argentina's future, and would be particularly important if war broke out again. To achieve these goals, Mosconi called for a "nationalistic economic organization" that would mobilize "the spirit of enterprise of national capital, which today is not invested or rests in mortgage loans." The army, Mosconi emphasized, should throw its technical and organizational ability behind the process of economic development.[57] Because petroleum was central to his vision of Argentina's industrial future, Mosconi believed that the state must control the oil industry. Absolutely opposed to additional foreign investment in Argentine petroleum, he condemned the multinational oil companies, the "mammoth vampires" that would stop at nothing, including bribery, to promote "sedition and treason" to gain control of the "black blood" they needed.[58] Critical both of Jersey Standard and of Royal Dutch Shell—he likened the former to a "hemp rope" and the latter to a "silk rope" because "both might hang us"—Mosconi nonetheless believed that Standard posed the more dangerous threat to the Argentine national interest.[59] He charged that the company's conduct in Salta and Jujuy demonstrated Standard's tradition of intervention in the host country's politics in order to gain its ends. To defend the national interest, Mosconi continued, Argentina must adopt a "closed door policy" against foreign oil investment and grant YPF "an integrated monopoly of all aspects of the industry: production, refining, transport, and distribution." Without such a monopoly, "it is difficult, indeed I would say impossible, for a state organization to emerge victorious in a commercial struggle with private corporations."[60]

Although Mosconi's support for an oil monopoly coincided with one aspect of Yrigoyen's petroleum policy, the YPF Director-General opposed the President's plan to expropriate existing private companies. Such a policy, Mosconi warned, was financially irresponsible and might involve Argentina in dangerous international complications, including a trade boycott.[61] Instead, Mosconi was willing to limit the foreign companies to their present holdings. On the matter of the future organization of the monopoly, Mosconi again broke sharply with Yrigoyen and opposed the President's statist solution. Well aware of YPF's failure to achieve the goals of the 1924–27 four-year plan, Mosconi by 1928 feared that a totally state-owned monopoly would lead to undercapitalization, excessive bureaucracy, and lagging petroleum production. To avoid these evils, he embraced a mixed-company plan along the lines of earlier proposals by the Unión Industrial Argentina and Alvear, but within the context of a state monopoly. Impressed by the British government's organization of the highly successful Anglo-Persian Oil Company, Mosconi proposed a "mixed monopoly" to be financed 51 percent by the government and 49 percent by private Argentine investors. As with Anglo-Persian, administration of the company would be in the hands of the private investors, although government members of the Board of Directors would retain veto power over major policy decisions. Thus constituted, the mixed monopoly "would be free from the slow administrative process characteristic of state organizations, and would convert into allies those groups who now opppose the state company."[62] Mosconi's program represented a sincere and realistic attempt to break the deadlock surrounding Argentine petroleum politics and to combine the basic goals of national ownership, maximum efficiency, and rapid growth. But Yrigoyen ignored the general's proposals and pushed ahead inflexibly to obtain Senate approval of the total state monopoly.

Mosconi Versus the Oil Companies

The serious plight of the Argentine petroleum industry during the late 1920's underscored Mosconi's arguments to reorganize YPF and to increase its capital base by allowing Argentine investors to participate. Consumption of all types of petroleum products was rising rapidly, and oil was replacing coal as the republic's prime energy source. Per capita consumption of coal during the 1926–30 period fell to about half that of the immediate prewar years, whereas consumption of petroleum products more than doubled between 1922 and 1930. Gasoline consumption led the petroleum-products boom, increasing over six times in the 1922–30 period.[63] Imports, however, supplied an ever-larger share of the Argentine market during the late 1920's, for domestic crude oil output could not keep pace with demand. Indeed, in 1930 output actually declined from the previous year, so that domestic production was able to supply only 42 percent of the demand, the smallest proportion since 1924 (see Table 4.3). The reasons for this flattening out of domestic production were that private companies, fearing expropriation, were declining to make new investments to increase output, and that YPF's production was stagnating.

The bright spots for the state firm were in Neuquén, where output rose steadily, and in Salta, where YPF rapidly pursued exploration and development of the oil fields it had acquired from Francisco Tobar (see Chapter 4). YPF began operations in Salta on January 1, 1928, and soon set up several drilling rigs and hired 350 workers. Ten wells were in production or being

TABLE 5.2. *YPF Oil Wells Completed and in Production, by Location, 1928–30*

Location	1928	1929	1930
Chubut			
New wells completed	140	165	165
Wells in production	604	808	946
Neuquén			
New wells completed	28	28	23
Wells in production	72	93	116
Salta			
New wells completed	1	2	2
Wells in production	1	3	5
TOTAL			
New wells completed	169	195	190
Wells in production	677	904	1,067

SOURCES: Argentina, [34], pp. 213, 249; Argentina, [12], p. 25.

sunk by the end of 1930, and a small refinery had been built to process part of the output for sale in the northern cities. However, most Salta crude, which was of very high quality, moved by rail to Santa Fe and then by barge to the La Plata refinery.[64] Although YPF's production in the west and north showed promising trends, output at the principal fields at Comodoro Rivadavia fell about 10 percent between 1928 and 1930. The disappointing production of the Chubut fields did not result from a lack of new wells, for YPF carried on an ambitious drilling program (see Table 5.2); rather, as we saw in Chapter 4, the water content of Comodoro Rivadavia crude continued to rise while per-well production fell.[65] This trend, which had begun in 1924, indicated that YPF's one large producing field was becoming depleted. To reverse the drop in output, YPF needed to undertake a major exploration program; yet because the state firm's profits were falling steadily it lacked the necessary capital to do so. Overdrilling and lack of exploration only partially explain YPF's stagnating production in the late 1920's, however; at a time of booming demand for oil, many of the state firm's storage tanks were full of unsold petroleum owing to YPF's inadequate distribution and sales network and its unwillingness to market through other distributors. Storage problems brought about by the backlog of oil production forced sporadic stoppages at both wells and refineries.[66]

Mosconi conceived a dramatic plan to reduce the unsold stocks, to attract public support for YPF, and, most fundamentally, to restrict the power of the international oil companies in Argentina. On August 1, 1929, suddenly and without warning, he cut the retail price of YPF gasoline. After a second reduction in November, the price had fallen by 17 percent. Previously, WICO and Anglo-Mexican, which together controlled over 70 percent of the republic's petroleum market, had set prices throughout the country.[67] But faced with YPF's reductions, the foreign-owned distributors also lowered their prices. Proud of his ability to force the multinationals onto the defensive, Mosconi claimed that "the English and North American oil trusts were definitively broken in our country." He was at least partially correct. For the next 30 years the Argentine government kept petroleum prices among the lowest in the world, and Mosconi's policy established an enduring image of YPF as a friend of the Argentine consumer.[68]

Mosconi planned his attack on the foreign companies at a time when world overproduction of petroleum was becoming a serious problem. The threat of oil shortages that had plagued the consuming nations before 1922 vanished as a stream of major discoveries in the United States, Venezuela, and the Middle East greatly increased supply during the 1920's. As prices dropped steadily in 1926 and 1927, a major new petroleum exporter, the Soviet Union, entered the world market. Soviet exports exceeded 3.2 million cubic meters in 1927, led to a price war in India, and posed a clear threat of a saturated petroleum market.[69] Troubled by the prospect of a worldwide price war, the principal international oil companies acted to limit production and stabilize prices. On September 17, 1928, representatives of Jersey Standard, Royal Dutch Shell, and Anglo-Persian gathered at Achnacarry, Henri Deterding's Scottish castle, ostensibly to shoot grouse. But they soon turned to more serious business and signed the famous Achnacarry (or "as is") worldwide price and marketing agreement. Under the terms of this pact, which at first remained secret, the companies agreed basically to form an international cartel in order to allocate shares of the world market and fix prices. Key provisions included maintaining each company's present volume of business in markets throughout the world, fixing future volumes accordingly, and setting a uniform international oil price based on the Texas Gulf Coast rate.[70]

While the petroleum multinationals were maneuvering to stabilize world prices, Mosconi was preparing to withstand a possible cutoff of imported oil by the foreign companies in retaliation for his planned price cuts. The key to Mosconi's plan was the Soviet Union. The Soviet trading firm

Iuyamtorg had opened an office in Buenos Aires in 1926, had incorporated under Argentine law the following year, and was expanding its operations rapidly. Between 1926 and 1928 it purchased an average of 19 million gold pesos of Argentine products yearly—mainly hides, *quebracho* extract, and wool. But Argentina imported much less from the USSR, and to redress this trade imbalance Iuyamtorg officials offered petroleum, one of the Soviet Union's major exports in the 1920's.[71] At an exposition of Soviet products in Buenos Aires in 1928, Iuyamtorg's brochure featured petroleum and assured its readers that Russia's reserves were "inexhaustible."[72] In 1929, YPF began to purchase gasoline from Iuyamtorg, and the Yrigoyen government undertook to negotiate a trade pact with the USSR. The key feature of the proposed pact from Mosconi's point of view was the provision that the Soviet Union would ship Argentina 268,750 cubic meters of gasoline annually for three years at a price of approximately 11 centavos per liter in exchange for a variety of agricultural products. Thus when Mosconi cut prices less than a year after the Achnacarry agreement, the international oil companies were outmaneuvered. Although they denounced the Russian gasoline imports as "dumping," they had little choice but to follow YPF's price reductions or lose the Argentine market.[73]

Mosconi implemented the YPF price cuts in a series of steps. In August 1929, YPF reduced prices of both gasoline and kerosene by two centavos per liter; in November gasoline fell two centavos more. Another cut on February 17, 1930, set a uniform gasoline price of 20 centavos per liter at all YPF outlets throughout Argentina. Because transporation and handling costs previously had kept prices much higher in the interior than in the capital, the net impact of Mosconi's decrees was to reduce gasoline prices in some localities as much as 34 percent.[74] YPF proudly proclaimed the large savings—estimated by Mosconi at 8.2 million pesos monthly—that Argentine consumers would enjoy from the price reductions. *The Review of the River Plate* estimated the savings at only about 30 million pesos annually, but agreed that the price cuts would provide significant "economy of transport" throughout the country.[75] Though the Argentine consumer gained, the privately owned oil companies' profit margins declined. Astra, for example, informed its stockholders in 1930 that despite a rise in international oil prices the preceding year, "the national market got no benefit from this advantage, owing to the sales at low prices effected by the Fiscal Administration." Troubled by YPF's competition and the threat of expropriation, several Argentine petroleum companies, including Astra, saw their stock prices tumble by about 50 percent on the Buenos Aires Exchange between 1929 and 1930.[76]

The price reductions were only one part of Mosconi's determined campaign to carve out a larger share of the Argentine petroleum market for YPF. The other part involved a vigorous expansion of the state company's distribution system. After the marketing arrangement signed with the Auger firm in 1925 expired in 1928, YPF negotiated to acquire the outlets Auger had established, and on May 1, 1929, assumed direct control over distribution and marketing through a new Sales Department.[77] Most of YPF's sales outlets were in the interior, however, which meant that in order to increase its share of the national market significantly the state company had to acquire a retail network in the capital, by far the largest petroleum market in the country. Municipal ordinances had virtually surrendered the Buenos Aires gasoline trade to WICO since 1915 (see Chapter 2), and as late as 1928 YPF had only 101 public sales outlets in a city of over two million people.[78] To overcome this initial disadvantage in Buenos Aires, Mosconi launched a fierce struggle to expand. YPF enjoyed the advantage of political support from the municipal government of Radical Intendant José Luis Cantilo, who in 1928 adopted a policy of transferring all expired concessions to the state firm and of granting it all licenses for gas pumps in public garages. Eventually, in February 1930, Cantilo ordered all agencies of the municipal government to utilize state petroleum products exclusively.[79] With Cantilo's help, YPF was able to quadruple its sales outlets in the capital between 1928 and 1930 (Table 5.3). But the foreign companies fought back to preserve their share of the market. Particularly in the rapidly growing Buenos Aires suburbs, WICO and Shell built modern service stations that contrasted with the lone gas pumps YPF was generally able to erect. The Radical Party press portrayed this commercial struggle in highly nationalistic terms. La Argentina, for example, published banner headlines exclaiming that "The West India Oil Company's Gas Pumps Symbolize the Dangerous Yankee Invasion" and urging readers to "Consume State Gasoline in Order to Liberate Yourself from the Extortion and Tutelage of Standard Oil." Despite such publicity and YPF's expansion, the private companies still controlled about 80 percent of the gasoline outlets in Argentina at the end of 1930.[80] Again, YPF's shortage of capital left it at a disadvantage; to compete effectively with the foreign sales networks, the state firm required more investment funds. But Mosconi's program for

TABLE 5.3. *Number of YPF Gasoline Pumps by Location, 1926—30*

Location	1926	1927	1928	1929	1930
Capital					
Public streets	18	26	79	156	272
Garages	0	19	22	128	258
Government agencies	1	20	25	31	33
SUBTOTAL: Capital	19	65	126	315	563
Interior	661	671	1,126	1,383	1,757
TOTAL	680	736	1,252	1,698	2,320

SOURCE: Argentina, [34], p. 379.

a mixed monopoly to raise capital from private sources still received no support from Yrigoyen, who continued to press his effort for a total state monopoly.

The Escalation of Petroleum-Policy Conflict

The main theater of conflict between the Yrigoyen government and the private oil companies remained the province of Salta, where the President's partisans quickly put Standard on the defensive. The Salta Radical Party had reunited to elect personalist Julio Cornejo as governor in the March 1928 provincial elections. Cornejo took office on May 1 and immediately announced his intention to drive Standard out of the province. Buoyed by Yrigoyen's promise to increase public works funding and to resume work on the Trans-Andean railway, Cornejo reversed the oil policy of his Conservative predecessor Joaquín Corbalán. The new governor reestablished the provincial petroleum reserve, cancelled all exploration permits granted since 1925 that did not conform to the national mining code, and ordered Standard to cease operations on lands disputed with YPF within the Tobar claim. The province also signed an agreement allowing YPF to explore for oil within a five-kilometer radius of Tobar's República Argentina well.[81] The effect of the governor's decrees was to force Standard to halt most of its Salta operations. In response, the company brought a lawsuit before the National Supreme Court in September 1928. Although the Court ordered the province to suspend Cornejo's decrees pending final judgment, Standard sharply cut back work in Salta and limited operations to completing drillings already under way and keeping existing wells in production.[82]

By late October 1929, when the Supreme Court began to hear the Standard Oil case, the legal conflict between the American company and the Salta government had drawn nationwide attention. The public galleries were packed when Silvio Bonardi, the young Radical attorney and protégé of Yrigoyen who represented the government of Salta, gave his emotional opening address. Placing the dispute in the context of national economic affairs, Bonardi argued that the true issues at stake were whether or not the federal government had the authority to direct the nation's economic policy, and whether or not Argentina would be subjected to "economic servitude" at the mercy of a foreign "trust." Reviewing the history of Standard's operations in Salta, he charged that the company had employed dummy representatives to hold its claims, had failed to carry out drilling operations within the prescribed time limit, and had committed numerous other legal violations.[83] Bonardi then called Francisco Tobar to testify. The 76-year-old Spaniard told the Court that under the Corbalán government "life was impossible for men who did not deliver themselves to the company," and that Corbalán's Minister of the Treasury had been on Standard's payroll.[84]

Rómulo S. Naón, one of the republic's best-known lawyers, presented Standard's case. Rejecting Tobar's testimony as "partial and self-interested," Naón delivered a meticulously constructed argument to the effect that Standard had acted within the limits of the law. Naón based his case on the mining decrees that Yrigoyen's "intervenor," or provisional governor, had established in 1918 when the federal government had intervened in Salta. Standard clearly had conducted its operation, argued Naón, within the terms of these decrees, which the provincial government had failed to repeal. Since Argentine constitutional history demonstrated that the decrees issued by federal intervenors were valid until repealed, Naón continued, Salta had no legal right to force Standard to cease its operations.[85] These arguments prevailed, and in June 1930 the Court issued a preliminary judgment in favor of Standard. Following the coup of September 1930, the Court suspended consideration of the matter until March 1932, when it declared Cornejo's 1928 decrees null and void.[86]

Its campaign against Standard stalled in the courts, the Radical Party resumed attempts to mobilize public opinion behind Yrigoyen's oil legislation. To put pressure on the Senate during its 1929 sessions, the party again organized university students and held public meetings in the principal Buenos Aires squares, including Plaza Italia and Plaza Constitución. Again General Baldrich played a prominent role. At a November 11 meeting in Plaza Once, the head of the Army Engineer Corps delivered a fiery speech before a large crowd and demanded that the Senate act.[87] Baldrich also entered the field of journalism when he assumed the editorship of the Buenos Aires Radical daily *La Argentina* late in 1929. To build circulation, the new editor relied on sensationalist reporting, heavy doses of patriotic rhetoric, and excellent sports pages. But most of all, *La Argentina* concentrated on sustained and passionate attacks against Standard. In his first signed editorial, Baldrich announced his support for "the nationalist campaign, for the question is whether to remain economically free or to submit to chains." Although he attacked the "filibustering, meddling, and degrading" influence of the "imperialist trusts," Baldrich carefully followed the lines of Radical economic policy when he emphasized that he supported "our approach to Europe . . . [whose] respectful loans do not come in warships."[88]

The Radical Party's campaign to organize public opinion against the oil companies received additional support from the Alianza Continental, an association formed in 1927 by the lawyer and intellectual Dr. Arturo Orzábal Quintana. Founded to combat foreign—particularly U.S.—economic imperialism, the Alianza proclaimed its intention to spark a movement throughout Latin America to protect the region's natural resources from foreign investors. In fact, however, this anti-imperialist group composed primarily of students and intellectuals concentrated almost entirely on the Argentine petroleum question. Orzábal Quintana, who had visited the USSR for two months in 1927 to study the Soviet oil industry, was convinced that Argentina should follow the Soviet example, evict the petroleum "trusts," and operate the oil industry itself.[89]

The Alianza's fascination with Soviet economic experiments and vehement anti-imperialist rhetoric led U.S. diplomats to conclude that it was a communist front funded by the Third International.[90] But this assumption was belied by the fact that YPF's bitterly anticommunist top officials were among the Alianza's major contributors. Mosconi's attitude toward the group is clear from his account of a meeting with its chief on January 15, 1929: "I had a long conference with the president of the Alianza Continental, Dr. Orzábal Quintana, in which we decided that it was indispensable to systematize the publicity campaign the group was carrying out in favor of oil nationalization and monopoly. We acted to create a great force of public opinion that would permanently defend Argentine sovereignty, which was more than ever threatened by the penetration of North

American capitalism." Asked by Mosconi to prepare a budget, Orzábal Quintana requested 5,000 pesos to finance a six-month campaign. The fund was raised by contributions from members of YPF's Board, and Mosconi himself contributed 200 pesos a month.[91] Throughout Argentina, the Alianza Continental propagandized over the radio, organized public meetings, and supplied speakers, who were often Orzábal Quintana himself and General Baldrich. The Alianza's publicists stressed Argentina's need for economic self-sufficiency, the significance of oil for the republic's economic development, and the "rapacious" penetration of the U.S. oil "trusts," specifically Standard.[92]

National attention riveted on the Senate when it convened in May for the 1929 sessions. But mired in partisan political conflict, the Upper House was unable even to begin considering oil policy until late in the year. During the first three months of sessions, all of its time was spent in debating the credentials of three new members elected the preceding year, Carlos Washington Lencinas of Mendoza, and Federico Cantoni and Carlos Porto of San Juan. Yrigoyen was determined to prevent the seating of these senators-elect, all of whom were leaders of populist movements that had split away from the Radical Party during the early 1920's. Deep political antipathy separated Yrigoyen from the Lencinas family of Mendoza and from Aldo and Federico Cantoni of San Juan. These two families had built powerful local political machines dedicated to social and labor reforms for the working classes as well as to provincial autonomy. Moreover, the Lencinas family was unalterably opposed to Yrigoyen's petroleum policy. After bitter and exhausting debate, the President eventually had his way when several Conservative senators joined the Radicals to prevent the seating of the three senators-elect.[93]

When the senators finally were able to turn to other business, they approached the petroleum issue with the utmost circumspection. On September 29, Molinari introduced a resolution to begin immediate consideration of the oil legislation passed by the Chamber of Deputies in 1928. But the opposition majority instead voted to delegate the entire oil question to a special study committee, which, once formed, requested information from Yrigoyen and opened an investigation into the affairs of YPF. When Yrigoyen ignored the committee's requests, the Senate took no further action before the 1929 sessions ended.[94] This attitude of caution earned the Upper House the applause of *La Prensa*, which urged a halt to "vehement rhetoric." Another prestigious voice that opposed the oil monopoly was that of Alejandro E. Bunge, who again emphasized his belief that the Radical Party's oil policy would actually impede Argentine economic independence. In a 1930 article, Bunge agreed that it was "absolutely necessary for us to free ourselves from the importation of foreign oil," but argued that the state monopoly never would be able to achieve that goal. Declaring that private investment in petroleum production was essential to satisfy national consumption, Bunge charged both Alvear and Yrigoyen with "ignorance," "malice," and "confusion." He warned that the oil issue had become "an instrument of street-level politicking," and that those responsible for making it one totally ignored the enormous economic issues at stake.[95]

The calls for restraint did not impress the Radical Party newspapers, which expressed outrage at the Senate's failure to act. *La Epoca* ran front-page headlines proclaiming that "The Senate is Against the Country." In a series of scathing editorials, *La Argentina* intimated that the Radical Party might employ violence to get its way: "The way things are going, we undoubtedly need a new Cromwell," it exclaimed, adding that the Senate was "provoking the people who . . . one day may pound at its doors with their work-calloused hands."[96]

The British Economic Offensive

The rhetoric of Yrigoyen's campaign against Standard Oil was highly nationalistic, but in fact the primary aim of the President's international economic policy seemed to be to strengthen the ties of dependence rather than to loosen them. With the sole exception of petroleum, the second Yrigoyen administration failed to launch any new economic policy initiatives. Rather than promoting industrial development and greater economic self-sufficiency, Yrigoyen formulated his economic policy along the lines the Sociedad Rural and the British embassy had been advocating since 1927. Throughout his second presidency, Yrigoyen's pro-British economic orientation was apparent. The British-owned railways, Paul Goodwin concludes, "were able to command an influence in the Casa Rosada comparable to the period of Conservative hegemony, prior to 1916."[97] The President warmly welcomed British bankers and applauded the decision of the Bank of London and South America to open additional Argentine branches. Ambassador Robertson reported with satisfaction Yrigoyen's remark that "we are and always have been entirely contented with British capital. . . . We do not wish to see it replaced by any other whencesoever it may come."

The British press and government regarded Yrigoyen very favorably. The London *Times*, for example, noted in an editorial that the President had been a "faithful friend" of Great Britain throughout his "long and distinguished career," and that his policies "cannot but be a cause of satisfaction in this country."[98]

Yrigoyen's determination to cooperate with the United Kingdom's trade policy became clear when he welcomed a high-ranking British economic mission led by Edgar Vincent, Viscount D'Abernon, in August 1929. To counter the Imperial Preference Movement, late in 1928 Molinari suggested to Robertson that an Anglo-Argentine economic conference might prove beneficial. Sensing an opportunity to reduce "the relentless competition of the Americans," as the *Times* put it, the British government enthusiastically accepted the suggestion, prepared an extensive agenda, and recruited prominent British business leaders to journey to Argentina.[99] The mission included Sir William Clare Lees, President of the Manchester Chamber of Commerce (representing the textile industry), Julian Pigott of the National Federation of Iron and Steel Manufacturers, and Argentine trade expert H. O. Chalkley, and it received a warm government welcome as soon as it disembarked at Buenos Aires. In sharp contrast to the frigid reception Yrigoyen had accorded visiting U.S. President-elect Herbert Hoover only eight months earlier, the government offered the British mission office space in the Casa Rosada, arranged a series of lavish receptions, and generally treated D'Abernon "like a visiting head of state."[100] The mission, however, came armed with a series of tough demands centered around threats to close the British market to Argentine goods in favor of products from the Dominions unless Argentina granted preferential treatment to British imports. "Argentina cannot survive without the British market," Robertson later noted. "We need the extra persuasion that a retaliatory tariff would provide."[101]

The two principal aims of the British mission were to sell equipment to the Argentine State Railways, which traditionally purchased locomotives, cars, and rails from less expensive Belgian and American suppliers, and to expand exports of "artificial silk," or rayon textiles, which the Argentine tariff laws classified in the high-duty category of silks. D'Abernon achieved the first aim by negotiating a reciprocal-trade agreement that obligated the Argentine government to purchase £9 million of railway equipment in return for British assurances to import an equivalent amount of Argentine farm products. Argentina's purchases were to be *additional*

19. President Yrigoyen and Viscount D'Abernon at the opening of the Sociedad Rural's 41st National Cattle Exposition, 1929. D'Abernon is the tall bearded man near the center of the photograph; to his left is Juan B. Fleitas, the Minister of Agriculture, and to his right is Federico Martínez de Hoz, President of the Sociedad Rural. Yrigoyen is just behind Martínez's right shoulder.

to existing trade, whereas Britain's were not. Although Yrigoyen cooperated fully in the reciprocal-trade negotiations, he hestitated at the British request to reclassify the import duty on "artificial silk"—perhaps because over 50 Argentine factories produced rayon fabrics to satisfy the fast-growing domestic demand. In return for a tariff cut, Yrigoyen asked D'Abernon to make a declaration that Britain would not levy import duties on Argentine meat, cereals, butter, and fruit. D'Abernon referred this matter to London, but Yrigoyen in the meantime, to show his friendly intentions, decreed a 50 percent reduction on duties on artificial silk goods imported from the United Kingdom. After the British government failed to give Yrigoyen the trade assurances he desired, and after France lodged vigorous protests against the discriminatory tariff reduction, the President refused to promulgate the artificial silk decree, much to the dismay of British importers in Buenos Aires.[102]

Despite the mission's failure to achieve the tariff reduction, London considered the D'Abernon agreements a significant victory. As Robertson exulted in 1930, Britain gained £9 million in new orders, and "in return we undertook to buy only a very small fraction of what we inevitably take. . . . *In fact, we obtained something for nothing.*"[103] In Buenos Aires, the D'Abernon pact aroused widespread criticism. "An agreement that cannot be rationally explained is morally dead," observed *La Prensa*. Enraged at Yrigoyen's willingness to sacrifice the textile industry to exporting interests, the Unión Industrial Argentina sharply attacked the President's policy. In the Chamber of Deputies, opposition to the reciprocal trade agreement was intense, but the Radical majority approved it on December 12, 1929.[104] However, the agreement never went into effect because the Senate, which postponed consideration of it in 1929, was unable to convene in 1930. Nonetheless, the pact demonstrated that the government, despite protestations of economic nationalism, was determined to maintain Argentine economic policy squarely on its traditional export-oriented basis and would abandon the principles of multilateral free trade in order to keep the British connection intact.

Further evidence of the Yrigoyen government's favorable attitude toward British business emerged in May 1930 when the President compromised his statist petroleum policy behind the scenes and conciliated the agents of Royal Dutch Shell. Diadema Argentina, a Shell subsidiary, possessed oil concessions in Chubut that were liable to expropriation under Yrigoyen's oil legislation, and another Shell subsidiary, the Anglo-Mexican Petroleum Company, operated a lucrative importing and distribution business in Argentina. Diadema had for years wanted to construct a modern oil refinery to process Anglo-Mexican imported crude, thus enabling Anglo-Mexican to escape the tariff on imported gasoline and putting the company on an equal footing with Standard Oil's subsidiary WICO, which long had operated refineries in Argentina. Diadema may also have hoped eventually to process its own production in such a refinery. In any case, Alvear had granted Diadema a construction permit just before leaving office, and the company had imported at least £300,000 of refinery equipment when Yrigoyen rescinded the permit and ordered construction to stop. In September 1929, Diadema's manager had an interview with the President, who emphasized that the pending oil legislation and the Standard Oil case then before the Supreme Court made an early decision impossible.[105] But the British Foreign Office instructed Ambassador Robertson to "take any unofficial steps which you may consider feasible to induce the Argentine government to reverse their present attitudes." Robertson soon did so, through a skillful maneuver: aware that Yrigoyen had recently approved the establishment of a new Petroleum Training Institute in the University of Buenos Aires, a facility YPF long had desired, the ambassador encouraged Diadema to offer the President on-the-job training for the Institute's students at the projected refinery. A few months later, Yrigoyen reversed himself and gave his approval for construction to resume.[106]

The Onset of the Great Depression

Although negotiations with British economic interests absorbed much of Yrigoyen's time late in 1929, the President was also busy preparing his party's strategy for the biennial elections to the Chamber of Deputies scheduled for March 1930. The Radical Party's strategy was to emphasize the petroleum issue, to win a "plebiscitary mandate" at the polls, and to use this victory to pressure the Senate to approve the petroleum legislation. During the election campaign, Radical politicians pointed to YPF's recent price reductions as evidence of the benefits the Argentine public would receive once Yrigoyen's nationalization and monopoly proposals became law.[107] The Radicals also portrayed Standard Oil and the Senate as partners in an unscrupulous conspiracy to prevent the Argentine people from enjoying the fruits of a nationalistic oil policy. For weeks prior to the elections, La Argentina ran banner headlines proclaiming that "Foreign Gold is Looking for Votes," "Standard Oil is Bribing Consciences," and "Ten Thousand Million Dollars Are at Stake!"[108] Several issues of the paper reviewed the history of oil in Salta—including the 1926 murder affair—and portrayed Standard's role in the darkest terms. Another prominent theme was the 1920 assassination of Mexican President Carranza, which the paper's editor, General Baldrich, blamed on Standard. Similar intervention by "filibusters who wish to convert the Republic into a Yankee colony" awaited Argentina if Standard remained in operation, Baldrich warned. "The future of the fatherland" was in the balance, but the Senate majority acted like "simple marionettes" who were "bribed by the Robbing Oil Company of the U.S.A." Radical street orators spoke of "direct action" against Senators who were "betraying the fatherland," and La Epoca raised the proposition that the Senate should be abolished.[109]

To fortify its argument that Standard Oil was taking advantage of the Senate's failure to approve the oil laws by carrying out a systematic take-

TABLE 5.4. *The Popular Vote in the National Elections of 1928 and 1930*

Party	1928	1930
Unión Cívica Radical	839,234	623,765
All opposition parties	536,908	614,336
RADICAL PLURALITY	302,326	9,429

SOURCE: Etchepareborda, "La segunda presidencia de Hipólito Yrigoyen," p. 363.

over of the Argentine petroleum industry, *La Argentina* devoted massive coverage to the Bolivian government's request to build a pipeline to Argentine tidewater in order to ship out petroleum produced in Standard's southeastern Bolivian fields. Standard had expressed interest in a pipeline to a deep-water port on the Paraná as early as 1922, and in 1925 one of the company's attorneys had broached the idea to Agriculture Minister Le Bretón and President Alvear, who rejected it and instead proposed shipping Bolivian oil through Argentina by railway. The Bolivian government regarded this plan with disfavor, partly because of the high costs involved and partly because it feared that reliance on railway links with Argentina would stimulate separatist tendencies in the oil-producing Santa Cruz region.[110]

The issue emerged again in 1929 when J. M. Escalier, Bolivia's minister in Buenos Aires, began conversations with YPF and the Ministry of Foreign Affairs to discuss construction of a pipeline that Standard would finance from Yacuiba on the Bolivian frontier to Formosa, Rosario, or Zárate on the Paraná. In statements to the press, Escalier emphasized that Bolivian oil production would stagnate at a low level for lack of an outlet if Argentina failed to grant permission for such a pipeline. However, the Bolivian proposal generated no support in the Argentine government, and Mosconi rejected the idea in December 1929.[111] Nonetheless, shortly before the election *La Argentina* seized on the projected pipeline as an issue worthy of major editorial concern. Bolivia, "a Yankee domain" according to the Radical paper, "serves the interests of the Standard Oil Company," and the pipeline would only further "Yankee intervention" in Argentina. The pipeline might well lead to "dumping" and "ruinous competition" against Argentine petroleum, the paper continued, and hence ought to be rejected.[112] When the Bolivian government subsequently decided to drop the request and postpone a special diplomatic mission it had planned to send to Buenos Aires, *La Argentina* took credit for alerting public opinion to the dangers of the project.[113]

The Radical Party's reliance on the petroleum issue failed in the 1930 elections, and Argentina's voters dealt a sharp blow to the Yrigoyen government. Where Radicals had outpolled the combined opposition by nearly 300,000 votes in 1928, they held only a bare majority of 9,400 in 1930 (see Table 5.4). Although the party gained three seats in the Chamber, its miserable showing in the capital, where it took third in the popular vote behind the two Socialist parties, indicated that the government's urban support was fast eroding in reaction to the administration's almost complete failure to combat the serious effects of the world depression on the Argentine economy. Yrigoyen's oil policies may well have evoked sympathy among the urban masses, but the financial crisis and the government's policy drift were matters of more immediate concern.[114]

The bloom of prosperity had begun to fade in Argentina in mid-1929. Aggravated by the failure of the wheat crop, the value of exports plunged. But because imports fell more slowly (Table 5.5), a serious international payments crisis soon developed. As gold poured out of the country during 1929, the domestic currency supply, which was tied to gold stocks, began to fall, and to avoid a complete economic collapse Yrigoyen ordered Argentina off the gold standard and closed the Conversion Office on December 16, 1929. But this maneuver did not save Argentina from deepening economic crisis in 1930. While the peso fell about 20 percent against the

TABLE 5.5. *The Impact of the Depression on the Argentine Economy: Indexes of Leading Economic Indicators, 1929–33*
(1928 = 100)

Economic indicator	1929	1930	1931	1932	1933
Exports	90.4	58.1	60.7	53.7	46.7
Imports	103.0	88.3	61.7	43.9	47.2
New gross fixed investment	114.4	96.3	58.9	42.0	n.a.
Real wages	99.0	91.0	98.0	104.0	96.0
Visible gold stocks	71.9	67.9	41.2	40.4	38.8
Gov't revenues	109.0	101.5	105.1	123.7	n.a.
Gov't expenditures	112.0	123.7	100.0	95.1	n.a.

SOURCES: For exports and imports, McCrea, Van Metre, and Eder, pp. 64, 72, and Díaz Alejandro, pp. 475–76, 461; for new gross fixed investment, Di Tella and Zymelman, p. 401; for real wages, Di Tella and Zymelman, pp. 399, 441; for gold stocks, Vernon L. Phelps, p. 263; and for government revenues and expenditures, Peters, p. 73.

dollar and the principal European currencies, the price of imports rose and total imports declined rapidly. Despite the fact that the drop in imports in turn reduced the government's customs revenues, Yrigoyen kept spending high, partially to maintain the flow of patronage. But this attempt to bolster the Radical machine proved insufficient in a year when the real wages of industrial workers declined by nearly 10 percent and when unemployment was spreading. Demoralized by Yrigoyen's failure to provide sufficient patronage, Radical ward bosses were unable to mobilize the vote and began to mutter criticisms of the President. Moreover, to cover its huge budget deficit, the government borrowed extensively, which restricted credit and alienated the landed elite. The depression, then, not only staggered the Argentine economy; it also seriously weakened the Radical political machine.[115]

While the depression was tightening its hold, the Yrigoyen government foundered in administrative paralysis. As in his first term, the President refused to delegate authority to his ministers, attempted to control the entire bureaucratic apparatus, and personally signed all government transactions. But Yrigoyen was no longer able to keep up with this extraordinary burden. Although he seldom missed a day at the Casa Rosada, "hardly showed signs of advancing age, and at times was remarkably youthful in his manner," as a British diplomat observed, his pace had slowed since 1922. Because effectively only Yrigoyen made decisions, his anterooms filled to overflowing with favor-seekers, businessmen, and politicians. All awaited the President's signature, but few received it and unfinished business accumulated rapidly on Yrigoyen's desk.[116] Nonetheless, Yrigoyen found time to devote to trivial matters: he made dramatic midnight appearances in his dressing gown to visit the victims of accidents, and conducted a personal vendetta against dancer Josephine Baker, whose nightclub acts scandalized him.[117]

The Yrigoyen government's legislative record matched its dismal administrative performance. Preoccupied with the petroleum issue and with engineering Radical electoral victories in the provinces, the government gave low priority to legislation to combat the country's deepening economic crisis. Although the administration did succeed in passing two important labor laws, which established the 48-hour week for urban workers and tightened enforcement of notoriously abused existing labor legislation, Yrigoyen adopted a cautious stance on most other policy issues.[118] His attitude toward rural development is illustrative: although agricultural economists, the press, and the Sociedad Rural all agreed that construction of a network of grain elevators was essential in order to rationalize Argentina's archaic agricultural storage and marketing system, Yrigoyen put forward no legislation on the subject.[119] In a 1930 memorandum to the President, the Sociedad Rural and a number of other producers' associations condemned the government's inaction on the storage and marketing question and emphasized that structural inefficiencies at a time of falling farm prices forced Argentine producers to sell crops unprofitably.[120] Although the government did back proposals to create an Agrarian Bank and to enforce existing legislation protecting tenant farmers from arbitrary exploitation and expulsion, the Radical Party's determination to give top priority to provincial political affairs prevented any congressional action on these proposals during the 1930 sessions.[121]

Isolated from the agrarian sector, the administration also alienated industrialists. Yrigoyen ignored the Unión Industrial Argentina, which urged the government to combat the rising rates of bankruptcy and unemployment with higher tariffs, vocational education, and industrial credit. In response—and particularly after the D'Abernon pact—the UIA became openly hostile to the government, launched a strong press and radio campaign against Yrigoyen's trade policy, and applauded the President's overthrow in the 1930 coup.[122]

The Breakdown of Political Legitimacy

Rather than attempting to mend his crumbling political fences, Yrigoyen directed his attention to the composition of the Senate. Determined to obtain his petroleum legislation, the President rejected compromises such as the Mosconi proposal and instead carried out a policy of provincial interventions that many political observers, at that time and since, have interpreted as aiming toward establishing a Radical majority in the Upper House. Yrigoyen's intransigence on the petroleum issue and his deliberate abuse of his intervention powers represented a direct presidential challenge to the federalist system and the constitutional rights of the provinces. The elites and the opposition parties responded to Yrigoyen's interventions with accusations that he was violating fundamental constitutional norms and that he planned to impose a Radical dictatorship in order not only to enact the state oil monopoly but, more importantly, to lead "a war against the social structure" and "an attack against one of its powerful bulwarks: the right of private property."[123] The President's petroleum policy, in other words, contributed to a serious legitimacy crisis and to the breakdown of the Argentine political system.

The opposition's fear that Yrigoyen aimed to become a "virtual dictator," as British diplomat Millington-Drake put it, increased sharply after April 1929, when the President decreed a highly controversial intervention in Corrientes province. Because Yrigoyen ordered this intervention after the Senate had rejected it, the President's critics charged that he had violated the Constitution. After listening to angry speeches comparing the Yrigoyen government with dictatorial regimes abroad, the Senate passed a resolution condemning the government for subversion of fundamental constitutional principles. Following the Corrientes intervention, the Independent Socialist Party (the Radicals' erstwhile ally in the 1927 petroleum debates) formally denounced Yrigoyen's government and claimed it was moving toward tyranny.[124]

The situation in the provinces of Mendoza and San Juan confirmed the worst fears of the political opposition. After violent incidents in the two western provinces, Congress had ordered federal intervention in them shortly before Yrigoyen returned to office in 1928. Yrigoyen's administration of the interventions, however, made it clear that the President aimed not to restore democratic electoral procedures but to prevent them. The President, in fact, ordered his provisional governors to prepare Radical victories in the March 1930 congressional elections. His intervenors arbitrarily jailed leaders of the opposition, seized thousands of voter registration cards, closed hostile newspapers, stuffed ballot boxes, and generally terrorized the followers of Lencinas and Cantoni. Argentina was stunned when an assassin killed Carlos Washington Lencinas, regarded as a hero by the Mendoza masses, at a political rally late in 1929. Although the motive behind the assassination never was clarified, outside commentators linked the Radical Party and the President himself to the act.[125]

Yrigoyen's vendetta against *lencinismo* and *cantonismo* led to a complete breakdown of Congress in 1930. Although the Radicals' fraudulent electoral tactics enabled them to win the March congressional elections in both provinces, two *cantonistas* and one *lencinista* were elected to the Chamber of Deputies on minority lists. But on Yrigoyen's instructions, when the Chamber met in May to open its annual sessions the Radical majority refused to seat the three western deputies-elect and another deputy-elect from Buenos Aires province. Debate over the credentials of these four, punctuated by angry charges of majority tyranny, totally absorbed over three months of the Chamber's time. Eventually Yrigoyen had his way and the Chamber voted to exclude them, but this political blood feud deeply discredited the Radical government and the Argentine political process. While Argentina staggered under the impact of the depression, the Chamber transacted no business other than the credentials affair in 1930, and the Senate, unable to convene until the Lower House had constituted itself, held no sessions at all. In Mendoza and San Juan, Yrigoyen's partisans, who needed to control the provincial legislatures in order to elect Radicals to the national Senate, were prepared to triumph in the provincial legislative elections scheduled for September 7, 1930. These elections, along with other forthcoming ones, by 1931 would have enabled the Radicals to increase their seats in the Senate to 13, or possibly 14—still two short of a majority. Additional provincial interventions would have given Yrigoyen his long-sought Senate majority.[126] But on September 6, 1930, the day before the San Juan and Mendoza elections, General José F. Uriburu, a patriarch of one of Salta's elite families, deposed Yrigoyen in a quick and almost bloodless coup.

Petroleum and the 1930 Coup

Soon after the events of September 6 it became a commonplace in Buenos Aires that Yrigoyen's overthrow "smelled of oil." During the 1930's, Raúl Scalabrini Ortiz, whose anti-imperialist revision of Argentine history attracted wide attention throughout the republic, argued that Standard Oil had played a central role in arranging the 1930 coup. The conspiracy thesis contends that the oil companies plotted with Uriburu to remove Yrigoyen, establish a military dictatorship, and cancel the projected nationalization, and it has continued to receive support from influential writers, many of them Radical Party militants, during the past 40 years.[127] When Carlos H. Perette, Vice-President in the Radical administration of Arturo Illia (1963–66), exclaimed that "33 years ago, a president fell because he defended Argentine petroleum," he was reflecting an interpretation that had become central to Argentine political mythology.[128]

The thesis that the oil companies were the prime cause of the 1930 coup relies on the "coincidence of personalities" argument, which emphasizes the close ties among the Salta elites, the oil companies, and the Uriburu government. As Oscar E. Alende, the Radical Governor of Buenos Aires province in 1958, explained it: "The exact facts are the following. The majority of the members of General Uriburu's cabinet in 1930 were linked to petroleum interests, whether of the American group (Standard Oil) or of the English group (Shell Mex). You can learn this from the corporate directories and from the publications of that era, which now are history."[129] It is known that several of Uriburu's ministers had served the foreign oil companies as legal advisers.[130] It may well also be true that Standard and other companies employed bribery or other means to influence the military to move against Yrigoyen. Multinational corporations certainly employed large-scale bribery abroad to gain their political ends long before the scandals of the mid-1970's finally publicized these tactics among the American public. Were the files of Jersey Standard open to the researcher, perhaps the issue of the oil companies' influence on the 1930 coup could be settled once and for all.

The historical evidence, however, suggests that the companies' meddling, if indeed it took place, was a peripheral factor and that the Yrigoyen government's campaign against Argentine federalism convinced the opposition that its alternatives were to permit one-party Radical rule or to remove the President from office. Johnson's model of economic nationalism points out an "evident connection" between a nationalistic economic policy in a developing country and "the propensity to establish one-party government."[131] The case of Yrigoyen's petroleum policy supports this observation. To establish the state oil monopoly that he had portrayed as a panacea to his middle-class power base, the President had to win control of the Senate and of those provincial governments where federalist sympathy was strong. Yrigoyen's intervention policy created a severe political split between the urban-oriented Radical Party and the interior provinces, and laid the foundations for the political crisis of 1930. The Radical chief's disregard of provincial rights and constitutional procedures shattered the confidence in the political system of the coastal elites and the elites of the oil provinces. The resulting legitimacy crisis, together with the government's paralysis and the impact of the depression, convinced Yrigoyen's opponents that democratic politics no longer functioned effectively. Yrigoyen's unwillingness to consider a compromise to settle the petroleum issue only intensified the political crisis. Alejandro Bunge and Enrique Mosconi, the most influential spokesmen of the 1920's for Argentine industrialization and economic independence, had proposed to integrate the national capitalist class into the state oil industry and thereby strengthen the future of YPF by expanding its capital base. But concerned primarily with strengthening his political movement, Yrigoyen ignored the mixed-company concept and tampered with the rules of the political game to achieve the state monopoly. This policy provoked the overthrow of Argentina's brief experiment with political liberalism.

CHAPTER VI

The Impact of Argentine Petroleum Nationalism, 1930-1977

Each December 13, when Argentines observe "Petroleum Day" to mark the 1907 discovery of oil at Comodoro Rivadavia, they also pay homage to the memory of Enrique Mosconi. The widespread veneration of the "petroleum general" is a direct reflection of the powerful influence petroleum nationalism wields in contemporary Argentina. Because Mosconi's ideas and actions mobilized powerful political support, YPF survived the 1930 coup, became the largest enterprise in Argentina, and emerged as the very symbol of national economic independence. Moreover, as the first state oil company in Latin America, YPF attracted attention throughout the region and became a model for national oil firms in several other republics. This final chapter sketches in broad strokes the development of YPF since 1930 and analyzes the impact Argentine petroleum nationalism has had on domestic policymaking as well as on the larger Latin American scene.

Oil and Politics: The 1930's

Argentine politics during the 1930's presented a sharp contrast to the liberalism of the preceding decade. General José F. Uriburu, who deposed Yrigoyen in 1930 and held power until February 1932, ruled dictatorially, and his conservative successors gained and held power through open electoral fraud. Persecuted under Uriburu and outraged at the manipulation of elections under the next President, General Agustín P. Justo, the Radical Party refused to participate in politics until 1935.

156

Enrique Mosconi was one of the most prominent casualties of this "infamous decade," as Argentine historians often call the 1930's. Uriburu, who regarded the prestigious "petroleum general" as a potential focus of opposition to his dictatorship, fired Mosconi from his post as Director-General of YPF on September 9, 1930. Firmly opposed to military rule, Mosconi refused Uriburu's request to collaborate with the new regime. The dictator responded by ordering Mosconi into exile in Europe. This exile was a severe blow to his health: shadowed everywhere by Uriburu's agents, Mosconi suffered deep emotional depression as well as physical illness. When he was allowed to return to Argentina after Justo took power, Mosconi encountered official insult. Rather than drawing on Mosconi's experience and administrative ability, Justo sentenced him to the obscure and powerless post of Director of the Army Fencing and Shooting Academy.[1]

After receiving this "decree of civil death," as biographer Raúl Larra has called it, Mosconi suffered a severe paralytic attack that left him bedridden for seven months. But the "petroleum general" fought back. He undertook a rehabilitation program and a course of exercise. Although confined to a wheelchair, he enlisted his sisters to serve as secretaries and in 1935 began to compose two books to publicize his ideology of petroleum nationalism throughout Argentina and Latin America. Mosconi published the first, *El petróleo argentino, 1922 –1930*, in 1936 and dedicated it to "the youth of Latin America," whom he urged "to accelerate the march toward our economic Junín and Ayachucho."[2] In 1938, he completed his second book, *Dichos y hechos, 1904 –1938*. Like his earlier work, this book emphasized that a state petroleum monopoly was essential to industrialization. Again in the Preface Mosconi appealed to "the youth of my fatherland" to lead the movement for "our integral independence."[3] These books attracted attention throughout the continent, and government oil officials from elsewhere in Latin America sought Mosconi's advice; nonetheless, the Buenos Aires regimes continued to ignore the former head of YPF. After completing *Dichos y hechos*, Mosconi's strength rapidly declined, and he died on June 4, 1940. But his state-monopoly thesis lived on to become the basic principle of Argentine petroleum nationalism.

By isolating Mosconi and removing the threat of nationalization, the Uriburu government created a favorable climate for the foreign oil companies, which expanded their operations dramatically and nearly doubled their output of crude oil during the two years following the 1930 coup.

Meanwhile, YPF suffered both administrative disorganization and financial stringency. Many observers had expected General Uriburu to curtail YPF's activities sharply because of his close ties to the Salta elites, and in fact YPF's outpost rose only 9 percent between 1930 and 1932. Nonetheless, as a military man, Uriburu retained a certain commitment to petroleum nationalism. He decreed enlargement of the 1924 state oil reserve to include the entire Argentine portion of Tierra del Fuego, and granted YPF authority to explore and produce oil in Salta and the other provinces—an authority that Mosconi had struggled to gain and that the governments of Alvear and Yrigoyen had been unable to give him.[4]

Standard Oil, which desired a contract with the Salta government to ensure future production, was unable to consolidate its position in that province during Uriburu's short term in the presidency. The pro-British exporting elite, led by the powerful General Justo, opposed Uriburu's plans for a corporate state and for protective tariffs and forced him to call national elections on November 6, 1931. Two days before the elections Standard and Salta signed a production contract, but Uriburu was obliged to cancel it a month later under heavy pressure from Justo, the President-elect. Opinion in Salta was outraged, for the province's economy was suffering a deep depression caused largely by Chile's suspension of all cattle imports from Argentina in 1930. Royalties from Standard's production were the only hope of a steady income for the province's exhausted treasury. The lameduck *salteño* in the president's chair, however, was unable to satisfy his province in the face of Justo's determined opposition to a Standard Oil contract.[5]

Although divided over the Standard contract, Uriburu and Justo both followed staunchly conservative social policies and refused to tolerate the organized labor movement in the government oil fields. Despite Mosconi's attempts to improve social conditions at Comodoro Rivadavia, the petroleum workers remained a repressed and volatile group. After the fall of Yrigoyen, the Argentine Communist Party began to organize the *Unión General de Obreros Petroleros* (UGOP), the first large oil workers' union to be founded since 1927. The Communists achieved rapid success, and by late 1931 the UGOP counted 3,200 members in the Comodoro Rivadavia region. After YPF and the private companies refused to negotiate with the new union, UGOP declared a general strike, which lasted for two weeks early in 1932. The government broke it with a powerful show of military force. The navy landed 2,000 marines, who jailed over 1,900 workers, deported over 1,000, and smashed the UGOP. Although unrest remained strong among the Comodoro Rivadavia workers, no unions legally existed in the Argentine oil fields from 1932 until Juan Perón came to power.[6]

In February 1932, General Justo assumed the presidency in the first of the series of rigged elections that characterized Argentine politics until 1943. Justo faced an exceedingly bleak international economic situation. The British government, sensitive to pressures from the powerful Imperial Preference Movement to limit food imports from outside the Empire, essentially gave Argentina the choice of losing the market for its major export —chilled beef—or accepting a set of tough economic concessions. Buenos Aires accepted Britain's demands in the 1933 Roca-Runciman Pact. In return for continued limited access to the British market, Argentina pledged to refrain from placing tariffs on duty-free goods (including coal), to reduce tariffs on other British goods to levels of 1930, to devote most of the sterling earned from exports to Britain to buying imports from that country, and to grant "benevolent treatment" to British capital invested in Argentine enterprises, especially railways and utilities. The Roca-Runciman Pact, whose terms were far more sweeping and favorable to the British than those of the 1929 D'Abernon agreement, sparked bitter public opposition in Argentina; but Justo's Congress ratified it in July 1933, and it shaped Argentine international economic policy until the Second World War.[7]

Argentina's increased economic dependence on the United Kingdom, together with the power structure of Argentine politics, placed severe limitations on Justo's petroleum policy. During his six years in power (1932–38), Justo attempted on the one hand to promote YPF at the expense of Standard, and on the other hand to reassure British investors, including those in the petroleum sector. Nationalist critics have charged that Justo sacrificed YPF to importing interests, but this judgment is too severe.[8] Although the Justo government opposed a state oil monopoly, it operated within the limitations placed on it to ensure the future of YPF during a period when foreign competition threatened to destroy the state company.

Although Justo strongly opposed Standard Oil's position in Salta, powerful provincial politicians forced the President to make concessions to the province's demands for a contract with the American company. Only a few months after Justo took office, Congress enacted Argentina's first national oil law. Although weaker than petroleum nationalists desired, the

1932 legislation reaffirmed YPF as the government's official petroleum enterprise, granted it the sole right to explore and produce on the state reserves, and gave it the right to import oil.[9] Bolstered by the President's support, YPF then urged the Salta government to sign a production contract. But the provincial legislature stalled, and in mid-1933 Salta defied the national government and signed a contract with Standard instead. Although in 1934 Salta also signed a production agreement with YPF, the American company had gained a favored position in the province. Salta was able to carry off this unprecedented challenge to the power of the national government primarily because a *salteño*, Robustiano Patrón Costas, perhaps the most powerful man in the province, had become the arbiter of Argentine politics during the Justo regime. Through careful organizational work and extensive ties among the elites of both the interior and the littoral provinces, Patrón Costas made himself the lynchpin and key architect of the *Concordancia*, the alliance of Conservatives, Independent Socialists, and dissident Radicals that supplied political backing to the Justo government. Where Uriburu had failed, Patrón Costas succeeded. Justo dared not intervene in the province or act to void the Salta contract for fear that Patrón Costas would withdraw his support from the government and cause a political crisis just when the President was negotiating the Roca-Runciman Pact with Britain.[10]

Justo let the Salta contract stand, but in 1934 he counterattacked to strengthen YPF's position in the provinces. In a decree, the President limited all private petroleum concessions to their existing boundaries; the rest of Argentina effectively became a huge YPF reserve. Congress ratified the decree in 1935, thus confirming YPF's right, originally granted by Uriburu, to operate in the provinces. Nonetheless, the provinces retained jurisdictional authority to negotiate oil concessions with the state enterprise, and all oil producers, public or private, had to pay a royalty of 12 percent of gross production to the provincial government (or to the federal government in the case of production in the national territories). YPF now expanded rapidly in the provinces, particularly Mendoza.[11] The private producers, however, reacted with bitter criticism against the enlarged state reserve.[12]

Prohibited from securing new concessions in Argentina, the foreign oil companies began to inundate the country with cheap imported petroleum. Imports of crude, at prices that undercut YPF, rose 100 percent between 1934 and 1935. In the press and in Congress, a chorus of protest arose charging that the foreign companies were employing "dumping" to drive YPF out of business.[13] Justo responded in 1936 with decrees that placed a ceiling on petroleum imports and that apportioned shares of both the import trade and the petroleum-product market among YPF, Shell, Standard, and the smaller private companies. These agreements, which remained in force until 1947, reserved about 50 percent of the national gasoline market for the state firm. Justo's strategy to protect YPF by dividing the market antagonized petroleum nationalists, who argued that the President's policy amounted to capitulation to the importers. Socialist Deputy Julio V. González, who led a congressional investigation of the agreements, called them "a death sentence against YPF." The correct course of action, according to the nationalists, would have been to expropriate the private oil companies for conspiring to destroy competition, a violation of Argentina's 1923 antitrust law. Despite these protests that Justo's policy unnecessarily favored importers, the foreign oil companies viewed the President's marketing agreements with dismay. "We are actually being squeezed at both ends," remarked one American oilman. The 1935 reserve law "restricted . . . all future efforts to increase our production of crude by obtaining new concessions, while now an attempt is being made to limit our sales."[14]

Standard Oil encountered far more hostility from the Anglophile Justo government than did the British petroleum companies. Although Standard had gained a production contract in Salta, Justo maintained a constant legal battle to prevent the expansion of the American company, whose position in Argentine oil production deteriorated rapidly. In 1936, Justo rejected Standard's petitions to combine its subsidiary companies and to increase its capitalization. In the same year, government investigators charged the company with espionage, unfair competition, and illegal importing of oil from Bolivia. These accusations found a receptive audience in the Argentine public, which had been deluged by charges that Standard had promoted the savage Chaco War that ravaged neighboring Bolivia and Paraguay from 1932 to 1935. As during the late 1920's, university students demonstrated to demand expropriation of the company, and General Baldrich, still active in the cause of petroleum nationalism, published widely read attacks against the American multinational. In 1937, Standard offered to sell its Argentine holdings to the government, but Congress failed to consider the proposal and the company continued operations. Its production, however, dropped sharply after 1934, so that Shell's subsidiary Diadema Argentina became the largest foreign oil producer in the country and expanded production steadily at its Comodoro Rivadavia concessions.[15] The negative public image that had surrounded Standard for better than two decades, combined with the government's pro-British economic orientation, relegated the American company to the status of an official pariah in Argentina.

Perón and Petroleum Nationalism

Although YPF expanded rapidly on the enlarged government reserve and more than doubled its output in the last half of the 1930's, the state enterprise was unprepared to meet the drastic energy crisis that struck Argentina during the Second World War. As during 1914–18, coal became scarce; but now, too, oil imports fell sharply, from 2.1 million cubic meters in 1939 to 492,000 in 1943. Although YPF's production grew 51 percent between 1939 and 1945, output on the private companies' limited concessions stagnated, with the result that national petroleum production rose only 23 percent during the war. In a repeat of the First World War crisis, the state firm faced not only a severe shortage of capital but also the virtual impossibility of acquiring machinery and equipment from abroad. Mosconi had established a policy of purchasing as much oil-industry equipment as possible from Argentine producers, and though the state's orders had stimulated the expansion of the national machinery industry, domestic factories remained vulnerable to wartime fuel and raw-materials shortages and were unable to supply YPF adequately. To meet the energy shortage, draconian rationing policies allocated available oil supplies, and fuel-starved Argentine industries again burned corn, wheat, and even peanut shells and rice husks. In 1944, the worst year of the fuel shortage, such emergency sources provided 31.4 percent of the republic's total energy consumption.[16]

Despite the energy crisis, Argentina's light industries grew substantially during the Second World War. Cut off from traditional sources of foreign exchange in the 1930's, Argentina had begun a process of import substitution to supply a vast array of consumer products previously purchased abroad. By 1943, for the first time in the country's history, manufacturing and construction provided a larger share of the Gross Domestic Product (GDP) than agriculture and livestock combined. Employment in the new manufacturing industries, which centered in the Buenos Aires region, grew impressively—from 890,000 in the 1925–29 period to 1,310,000 in 1940–44.[17] Massive internal migration out of the interior, which reached a net annual total of 117,000 between 1943 and 1947, accompanied the industrial boom.[18]

In the midst of this rapid social and economic change, Colonel Juan Perón began his dramatic rise to power, which culminated in the 1946 presidential elections. Unlike earlier military presidents, Perón emphasized social justice and dignity for the Argentine workers. After becoming head of the Secretariat (later the Ministry) of Labor in 1943, Perón followed a policy of redistributing income and promoting unionization that attracted massive political support among both recent migrants and older segments of the Argentine working class.[19] In the oil fields, Perón permitted unionization for the first time since 1932. The Federación de Sindicatos Unidos Petroleros del Estado (SUPE), a nationwide, government-backed oil workers' union, achieved major wage raises and improvements in labor and living conditions for its members. In return, SUPE lent its support to petroleum nationalism—the first time since the foundation of the state oil industry that the petroleum workers embraced the cause of YPF. Perón summarized his determination to integrate the oil workers and the rest of Argentine labor into his regime when he told a SUPE gathering in 1949 that "I want production without exploitation. We want to intensify our oil-field operations but we do not want to exploit our workers."[20] Nationally, the labor unions became the major power base of the first Perón regime (1946–55). After Perón's overthrow, the unions used strikes and, when possible, political mobilization to defend the economic and social benefits they had gained under him. Although Peronism made a profound impact on many aspects of Argentine life, its most enduring legacy was the politicization of the working masses.[21]

Perón's ambitious plans for the Argentine economy included increased industrialization, nationalization of foreign transportation and utility investments, and reduction in the country's economic dependence. Nevertheless, Argentina's economic structure did not change significantly by 1955. Industrial production increased, but the growth was mostly in light consumer industries rather than in heavy industries such as steel or petrochemicals. Argentina still relied on nonindustrial exports as the motive force of its economic growth. During the boom years of 1945–48, when the value of exports jumped from 700 to 1,500 million dollars, the nation prospered and the annual growth rate of the GDP was about 10 percent. But after 1949, the terms of trade moved against Argentina, acute foreign-exchange shortages appeared, and per capita income declined.[22] Within this economic context, Perón attempted to devise a petroleum policy that would satisfy the large and vocal bloc of economic nationalists and that also would provide sufficient energy for the growing economy. But beset by Argentina's unstable international economic position, Perón was unable to pursue a consistent energy program.

During the first years of his presidency, Perón swung toward the state-monopoly position. Riding the crest of the export boom, the Argentine government by 1947 was engaged in massive and expensive expropriations of foreign-owned railways and utilities. Intense pressure was put on Perón to expropriate the foreign oil companies as well by petroleum nationalists like Julio González, who compared them to "a fifth column hidden in our country."[23] On December 13, 1947, Petroleum Day, Perón appeared to accept the nationalist position when he declared that "Argentine oil policy . . . must be based on the same principles as the rest of our economic policy: absolute preservation of Argentine sovereignty over our subsoil wealth and rational scientific operation carried out by the State." Two years later, Perón's new Constitution, which declared all mineral resources the inalienable property of the nation, gave the central government jurisdiction over all oil concessions for the first time in Argentine history.[24] Yet despite his rhetorical commitment to petroleum nationalism, Perón did not expropriate the private oil companies, whose production was dropping steadily. Instead, the President staked his energy policy on launching a national coal industry, utilizing Argentina's hitherto neglected natural gas resources, and intensifying YPF's production. The 1947 five-year plan envisioned a 50 percent increase in government oil production by 1951.[25]

But YPF was unable to keep pace with the rapidly industrializing economy's thirst for petroleum. When Perón assumed the presidency in 1946, most of the state oil company's capital plant was worn out or antiquated, and imports of machinery and equipment were extremely difficult to arrange in the tight postwar market. Determined to oppose Perón because of his alleged pro-Fascist sympathies, the United States adopted a policy emphasizing, in the words of the Acting Secretary of State, that "it is essential not to permit the expansion of Argentine heavy industry." Accordingly, the United States placed severe restrictions on the export to Argentina of oil drilling equipment, refinery apparatus, and repair parts for the petroleum industry. This policy substantially slowed the pace of renovation at YPF.[26] When the United States eased its restrictions in 1948, Argentina already faced the end of its postwar economic boom and was suffering foreign-exchange shortages that impeded the importing of sufficient quantities of vital capital goods—not only for YPF but for the entire economy. Throughout the late 1940's, the Argentine machinery industry remained unable to supply the drilling rigs and other specialized tools that were essential to expanding production and exploration.[27]

These severe postwar equipment shortages, along with Perón's cumbersome bureaucracy, forced YPF to restrict exploration and drilling to levels completely incompatible with the economy's rising demand for oil. In 1947, an average of only 33 drilling rigs worked full-time, and YPF completed 105 new wells—fewer than in 1925. As late as 1955 the state company possessed only 45 rigs, although it required at least 100. Production inched up 22 percent between 1947 and 1951, a period during which petroleum imports doubled to nearly $200 million yearly.[28] By 1955, Argentina's petroleum consumption reached 11.1 million cubic meters, which was more than double national production. Although natural gas output increased greatly in the early 1950's, the state coal industry remained in its infancy. Perón's energy policy, hobbled by his government's financial embarrassment since 1950, clearly had failed.

In May 1955, Perón dramatically abandoned his commitment to petroleum nationalism. Faced with a $244 million trade deficit that year, he concluded that Argentina could not afford to import ever-increasing quantities of petroleum and signed a provisional contract to grant a huge oil concession to the *Compañía California Argentina de Petróleo*, a subsidiary of Standard Oil of California (SOCAL). Industry observers considered the contract provisions very favorable, for they offered SOCAL a 40-year concession to explore and produce oil on a 50,000-square-kilometer area in Santa Cruz. YPF would buy the company's output at the Texas Gulf price, the government would receive 50 percent of the company's profits, and the company would be free to remit its net earnings to the parent firm. In a message to Congress, Perón urged the legislators to approve the contract on the grounds of Argentine petroleum self-sufficiency. He failed to mention the question of national ownership.[29]

The Standard Oil concession proposal ignited loud and bitter protest throughout Argentina. Perón's own party divided over it, whereas the opposition Radical Party found the issue a "godsend," as Arthur Whitaker has put it. The Radical National Committee, chaired by Arturo Frondizi, appealed to the public with a policy statement charging that "foreign capitalists would become the economic masters of Argentina." Adolfo

Silenzi de Stagni, a professor at the University of Buenos Aires Law School, published an extremely influential little book, *El petróleo argentino*, which examined the contract in detail, argued that it would prove economically disadvantageous to the nation, and concluded that Argentina would "find itself in the same situation as nineteenth-century China." The distinguished and highly respected Socialist politician and intellectual Alfredo L. Palacios condemned it as a "disgraceful contract." The Radicals, along with a chorus of other critics, warned that provisions allowing the company to build airstrips, roads, and other support facilities served as a mask for the installation of U.S. military bases in Argentina.[30]

Like his opponents, Perón attempted to defend his petroleum policy with nationalistic arguments. Portraying his quest for self-sufficiency as representing the true interest of Argentine nationalism, the embattled President lashed out at his detractors as "comic-opera nationalists" whose "stupid policies have hurt the country as much as the tricks of the colonialists." Perón's defenders also suggested that opponents of the concession were allied with oil-importing interests. But these arguments carried little weight in a country where petroleum nationalism virtually had identified government ownership of the oil industry with Argentine sovereignty for nearly 40 years.[31]

The Frondizi Oil Contracts

The Argentine Congress never had a chance to act on the SOCAL concession, for only four months after its announcement, a military coup unseated Juan Perón. The oil contracts had intensified the deepening alarm with which many military officers viewed the Perón government. Shortly after the coup, provisional president General Eduardo Lonardi told the Argentine people in a radio address that "the entire nation" had repudiated the "unacceptable arrangements" with SOCAL.[32] One year after Perón's overthrow, historian Arthur Whitaker, an expert on Argentina, wrote that "it would not be politically feasible for any Argentine government in the near future" to invite foreign petroleum investment, "least of all" from the United States.[33] But in 1958, newly inaugurated President Arturo Frondizi ignored this sage advice and again turned to U.S. oil companies to solve Argentina's economic woes. As was the case with Perón, this decision severely weakened Frondizi's power base. The tremendous political uproar that the 1958 contracts unleashed clearly illustrated the continuing strength of Argentine petroleum nationalism.

Frondizi's decision to rely on foreign oil investment opened him to charges of duplicity, for he had established a reputation as a thoroughly committed economic nationalist. A longtime Radical activist, as a young lawyer he had supported Hipólito Yrigoyen and begun a steady climb through the party ranks to a position of leadership during the Perón era. In 1954 he published a famous book, *Petróleo y política*, a history of the Argentine oil industry with a strongly anti-imperialist emphasis. The book made eloquent pleas to nationalize all "key industries" and to place the entire petroleum industry under a YPF monopoly. After a bitter Radical Party split in 1957, one faction, the Unión Cívica Radical Intransigente (UCRI), named Frondizi its presidential candidate for the 1958 presidential elections. The other faction adopted the name Unión Cívica Radical del Pueblo (UCRP). After a vigorous presidential campaign that emphasized economic nationalism and social justice for the working classes, Frondizi won a plurality of 45 percent of the popular vote and assumed the presidency in October 1958.[34]

Frondizi then shocked the nation by completely reversing the economic policies he had long and strenuously advocated. The extremely serious state of the Argentine economy formed the context in which Frondizi reached his decision to invite massive foreign investment. Since the late 1940's, the terms of international trade had not favored Argentina: prices for the country's agrarian exports did not keep pace with those for the industrial goods it needed to import, which contributed to chronic balance-of-payment deficits. Aggregate economic production stagnated, and the per capita GDP in 1957 was no larger than in 1949. Inflation reached 32 percent in 1958, and spiraled to 114 percent by 1959. Working-class purchasing power suffered severely, and average real family incomes did not reach the level of the postwar peak year of 1949 until the mid-1960's.[35]

The new President formulated an ambitious development plan that aimed to end Argentina's export dependence and to stimulate rapid and sustained economic growth. Massive foreign investment in heavy industry and petroleum, it was assumed, would propel Argentina toward Frondizi's goal of a modern, powerful, and independent nation. Oil held top priority in this development plan. In 1957, petroleum imports had cost the nation $272 million in scarce foreign exchange and had reached 62 percent of Argentina's petroleum consumption, up from 48 percent in 1946. According to Arturo Sábato, whom Frondizi named to administer YPF, total oil imports between 1951 and 1958 had cost $1.7 billion, a sum greater than the nation's total balance-of-payments deficit in the same period. Under these circumstances, Frondizi abandoned his state-monopoly principles and negotiated a series of contracts with U.S. oil companies.[36]

These contracts took several forms. Some were exploration and drilling contracts for wells that were to be turned over to YPF in exchange for payment according to the number of meters drilled. Fifty percent of the payments were to be in dollars. Others were production contracts, under which the companies sold their output to YPF at import-parity prices.[37] The agreement that the Frondizi government signed with Pan American Oil Co. (a subsidiary of Standard Oil of Indiana) was typical of this category: the company received a 30-year right to drill and produce on 4,000 square kilometers of land already explored by YPF in the Comodoro Rivadavia region, and YPF agreed to pay $10 per cubic meter for the oil, 60 percent in dollars. This price, much greater than the cost of production, enabled the company to reap annual profits of 100 percent or more, which were tax free and subject to no limits on repatriation.[38]

Frondizi's oil policy required new legislation to "nationalize" Argentina's oil deposits, however, since Perón's 1949 Constitution had been scrapped after the 1955 military coup and concession jurisdiction had reverted to the provinces. This was accomplished late in 1958, when despite serious opposition the government won approval of a new oil law transferring the provinces' rights to grant oil concessions to YPF.[39]

Attracted by the favorable terms the Frondizi government offered, nine oil companies invested about $200 million in Argentina between 1958 and 1963. The results were spectacular. Between 1959 and 1962 more new wells were drilled in Argentina than in the previous quarter century. Oil output rose from 5.7 million cubic meters in 1958 to 15.6 million in 1962, and Argentina approached self-sufficiency in crude production. Petroleum import costs declined from 22 percent of the nation's total import bill in 1957 to only 3 percent in 1963. Euphoria swept foreign investors in Argentina, for the country seemed "chosen," as Peter Odell put it, "to demonstrate the ability of international capitalism to secure the growth in wealth of the developing nations of the world." For the first time, the multinational oil companies had broken the united front that the state oil companies in Argentina, Brazil, and Chile long had maintained against foreign investment.[40]

But Frondizi's experiment with private oil investment soon encountered the hard political realities of Argentine petroleum nationalism. Led by the

UCRP, the faction of the Radical Party that had opposed Frondizi and his UCRI, a chorus of prominent Argentines condemned the contracts as scandalous giveaways to the foreign companies. Professor Silenzi de Stagni, who again launched a nationalist campaign, charged the government with "treason against the Fatherland." Silenzi and other critics claimed that the companies' large profit remittances more than negated the import savings increased production made possible, and that the dollar payments YPF had to make for the contractors' oil placed the state enterprise in financial jeopardy. Frondizi's method of negotiating the contracts—in secret and without the consent of Congress—received scathing criticism.[41] The President insisted that his policy did not compromise nationalism because Argentina was becoming self-sufficient and YPF retained ownership of the oil, but many Argentines would have agreed with the conservative economist and Latin American expert Simon G. Hanson, who wrote that the terms were "extremely prejudicial to the Argentine people."[42]

As had been true since the days of Yrigoyen, support for petroleum nationalism arose among the military, the bureaucracy, the middle classes, and the intellectuals. But in 1958, Argentina's potent organized labor movement also joined the nationalist coalition to protest Frondizi's oil policies. The oil-workers' union, SUPE, regarded the production contracts as the entering wedge of an attempt to drive YPF's 30,000 workers out of their jobs, and condemned the agreements as a threat to "national sovereignty, the integrity of YPF, and the interests of the workers."[43] To demonstrate its position, SUPE's Mendoza sector called a strike on November 1, 1958, that within days imperiled fuel supplies throughout Argentina. The strike also touched off a major national political crisis that dramatized the cause of the petroleum nationalists and exposed the weak political base of the Frondizi government. After SUPE's national leadership threatened to call a general petroleum work stoppage, Frondizi denounced the strike as a Peronist-communist subversive plot, called a state of siege, and unleashed the police and the army to repress the labor movement. Although the army hierarchy supported Frondizi, who jailed hundreds of union leaders, his own party split deeply over the government's response to the petroleum strike. After three weeks the strike ended, but only when Frondizi agreed to negotiate further with SUPE.[44]

Petroleum nationalists eventually received an opportunity to reverse Frondizi's oil policy. By 1962, the government was in serious trouble on both the economic and the political fronts. Frondizi's economic development plan had produced only mixed results: the balance of payments was positive, but industrial production had risen a mere 6 percent over the 1958 level; inflation remained serious; and in 1962 per capita GDP declined. Perón's following among the working classes remained strong, and the Peronist Party, legalized by Frondizi, swept important 1962 provincial gubernatorial elections. Shortly after these elections the military deposed the hapless President in a quick coup.[45]

After a year of military rule, presidential elections were held, from which the Peronists again were excluded. The results have been interpreted as showing that 70 percent of the votes cast were for candidates opposed to Frondizi's oil policy.[46] The new President, veteran Radical leader Arturo Illia, made petroleum nationalism the foundation stone of his government's policy. In keeping with his campaign promises, as his first major action, Illia decreed the Frondizi contracts null and void on the grounds that Congress never had approved them. YPF assumed control of the contractors' properties, and the Chamber of Deputies opened a full-scale investigation into the entire history of the agreements. The investigating committee's report, issued in 1964, claimed that the contracts were fraudulent concessions in disguise, charged the companies with bribery, and marshaled statistics to argue that profit outflows exceeded import savings. After the United States pressed his government to compensate the companies for the seized properties, Illia eventually agreed to negotiate compensation.[47]

Petroleum, the Army, and "National Security"

Bedeviled by continuing economic stagnation and bitter political disputes over the question of Peronist electoral participation, Illia lost power in another military coup in July 1966. The new President, General Juan Carlos Onganía, a hard-line conservative and authoritarian, again turned to foreign investment as the answer to Argentina's problems. This decision reflected profound ideological changes that were taking place among Argentine army officers. Since Perón's overthrow in 1955, no really consistent economic policy program had existed within the army. But since 1963, a group of military intellectuals centered around Major General Juan E. Gugliamelli had emerged, and during the Onganía presidency this group strongly influenced the education of Argentine officers. General Gugliamelli formulated for the army's various schools and academies a "Doctrine of National Security" that linked Argentina's security to the defeat of "internal subversion." To accomplish this goal, the military was to encourage social and economic development, which would require heavy industrialization, financed by foreign capital if necessary. Not all army officers agreed with Gugliamelli, and questions involving the status and future of the state-owned companies continued to divide the officer corps.[48] But the "National Security" doctrine, often called "developmentalism," was on the ascendant in the officer corps, and Onganía's civilian advisers urged policies that dovetailed with it.

Accordingly, the private petroleum sector again enjoyed government support. The Onganía government negotiated new contracts with Pan American Oil and Cities Service, the only two companies that had not yet concluded compensation agreements. A new petroleum law that permitted the government to grant private concessions in newly discovered oil zones, including the continental shelf, also was decreed. Petroleum nationalists responded bitterly. A group of labor union leaders, for example, denounced the Onganía oil law as invalid and declared that "the heritage of General Mosconi . . . survives with more vigor than ever among the working class."[49]

Encouraged by Onganía, whom the U.S. State Department expected to remain firmly in power, the private companies expanded both operations and output during the late 1960's. But the Argentine masses, highly politicized and still loyal to Perón, rebelled against Onganía's draconian policies of sharply reducing working-class real incomes. A series of riots that were virtual insurrections struck Córdoba and other industrial cities in 1969 and 1970. The riots severely damaged Onganía's image and programs, and in June 1970 the military forced his resignation.

The presidents who succeeded Onganía, General Marcelo Levingston (1970–71) and General Alejandro Lanusse (1971–73), concentrated on political matters—especially on combating Argentina's rapidly escalating guerrilla activity—and made no new initiatives on petroleum policy. Not until 1973, when Lanusse allowed elections that returned the Peronists to power, did oil become a matter of serious public debate again. Perón's hand-picked presidential candidate, Héctor Cámpora, adopted a strong nationalist position toward foreign investment during his brief term in office, which began on May 25. Then, unable to control political violence, Cámpora and his Vice President resigned on July 13 to enable Juan Perón to regain power through new elections. Perón and his third wife, María Estela (Isabel) Martínez, were elected President and Vice President of Argentina on September 23 and were inaugurated on October 12.

The political climate again grew hostile toward the petroleum multinationals, several of which continued to produce oil in Argentina. A series
of Peronist decrees considerably reduced the profitability of the foreign
companies' refining operations and gave YPF a monopoly over petroleum-
product marketing.[50] When Congress began to consider new oil legislation, the Peronist government proposed to nationalize all foreign oil refineries but to leave current concessions and production contracts intact; the
reunited Radical Party, however, true to its tradition, sponsored a much
tougher bill that called for an immediate government takeover of the entire
private petroleum sector.[51]

The shock waves of the world oil crisis delayed passage of a new round
of nationalistic petroleum legislation. Argentina again was becoming dependent on oil imports. The private contractors had ceased expanding production in 1971, and YPF was unable to raise its output enough to keep
pace with rapidly rising demand. The capital shortages that have plagued
YPF throughout its history had again become severe after the Illia annul-
ments, partially because of the burden of settling the canceled contracts.
A more important cause of the capital shortage was that YPF had been
receiving only about 31 percent of the sale price of the gasoline it produced;
the government took nearly all the rest in taxes. To increase its insufficient
profit margin, YPF proposed to raise prices to international levels, a policy
the government has opposed because of its inflationary impact. Given the
government's determination to finance itself through YPF on the one hand
and to keep Argentine fuel prices low on the other, YPF can finance major
expansion only through foreign borrowing (difficult to arrange) or through
cooperation with private firms.[52]

This capital shortage has restricted YPF's operations severely. During
the early 1970's, it was estimated that the state company needed to invest
the equivalent of at least $500 million annually to make Argentina self-
sufficient; but YPF was never able to muster more than about half that
amount. In terms of operations, the capital shortage meant that YPF was
able to drill only about 600 of the 800 new wells needed annually to reach
self-sufficiency. Moreover, money was not available to finance much exploration, so that proven reserves began to fall steadily. YPF's 1969 discovery of the rich Caimancito field in Jujuy had lifted proven reserves to
392 million cubic meters in 1970, but no other big finds were made until
1977.[53] The last year when crude production came close to satisfying
consumption was 1972, after which output dropped about 3 percent an-

TABLE 6.1. *Argentine Domestic Production and Imports of
Petroleum, Selected Years, 1930–74*

(Thousands of cubic meters)

Year	Total petroleum consumption[a]	Domestic production[b]		Imports[c]	
		Amount	Percent of total consumption	Amount	Percent of total consumption
1930	3,431	1,431	41.7%	2,000	58.3%
1935	3,873	2,273	58.7	1,600	41.3
1940	5,199	3,277	63.0	1,922	37.0
1950	9,002	3,730	41.4	5,272	58.6
1955	11,131	4,850	43.6	6,281	56.4
1960	14,330	10,153	70.9	4,177	29.1
1961	17,528	13,428	76.6	4,100	23.4
1962	18,714	15,614	83.4	3,100	16.6
1963	17,144	15,444	90.1	1,700	9.9
1964	18,143	15,943	87.9	2,200	12.1
1965	20,725	15,625	75.4	5,100	24.6
1966	22,056	16,656	75.5	5,400	24.5
1967	21,784	18,232	83.7	3,552	16.3
1968	22,132	19,951	90.1	2,181	9.9
1969	23,058	20,681	89.7	2,377	10.3
1970	26,071	22,798	87.4	3,273	12.6
1971	27,577	24,565	89.1	3,012	10.9
1972	26,985	25,193	93.3	1,792	6.7
1973	29,142	24,442	83.9	4,700	16.1
1974	28,774	24,274	84.4	4,500	15.6

SOURCES: Dorfman, p. 143; *International Petroleum Encyclopedia*, pp. 240, 292; Villar Araujo, p. 14; Zinser, "Alternative Means of Meeting Argentina's Petroleum Requirements," p. 190; "Argentina—Petroleum," *Bank of London & South America Review*, 9 (Apr. 1975), p. 201.
[a] Includes exports, which never exceeded 0.5 percent of consumption.
[b] Crude petroleum only.
[c] All petroleum products.

nually. Declining production forced Argentina to import more oil just when world crude prices exploded in 1973. As a result, petroleum imports again became a heavy burden on the balance of payments. In 1974, the oil import bill amounted to $455 million (20 percent of Argentina's total import bill), up from $115 million the year before.[54]

After Juan Perón died on July 1, 1974, his widow governed Argentina with an inept and ineffective hand until March 24, 1976, when the military again seized power. The "developmentalist" ideology that had emerged in the army a decade earlier now began to influence oil policy again. President Jorge Videla announced a new petroleum program, which he called "nationalism with objectives." Its target was to achieve self-sufficiency by

TABLE 6.2. *State and Private Petroleum Production in*
Argentina, 1930–58
(Thousands of cubic meters)

| | | State fields | | Private fields | |
Year	Total domestic production	All fields	Pct. of total domestic production	All fields	Pct. of total domestic production
1930	1,431	828	57.9%	603	42.1%
1931	1,862	874	46.9	988	53.1
1932	2,089	902	43.2	1,187	56.8
1933	2,177	922	42.3	1,255	57.7
1934	2,230	836	37.5	1,394	62.5
1935	2,273	944	41.5	1,329	58.5
1936	2,457	1,140	46.4	1,317	53.6
1937	2,600	1,262	48.5	1,338	51.5
1938	2,715	1,431	52.7	1,284	47.3
1939	2,959	1,625	54.9	1,334	45.1
1940	3,276	1,983	60.5	1,293	39.5
1941	3,500	2,227	63.6	1,273	36.4
1942	3,769	2,446	64.9	1,323	35.1
1943	3,948	2,633	66.7	1,315	33.3
1944	3,852	2,576	66.9	1,276	33.1
1945	3,638	2,457	67.5	1,181	32.5
1946	3,307	2,260	68.3	1,047	31.7
1947	3,473	2,426	69.8	1,047	30.2
1948	3,692	2,646	71.7	1,046	28.3
1949	3,591	2,580	71.8	1,011	28.2
1950	3,730	2,755	73.9	975	26.1
1951	3,889	2,958	76.1	931	23.9
1952	3,946	3,097	78.5	849	21.5
1953	4,532	3,711	81.9	821	18.1
1954	4,702	3,917	83.3	785	16.7
1955	4,850	4,067	83.9	783	16.1
1956	4,930	4,153	84.2	777	15.8
1957	5,398	4,656	86.3	742	13.7
1958	5,669	4,964	87.6	705	12.4

SOURCE: Instituto Argentino del Petróleo, *El petróleo en la República Argentina*, p. 5.

1980 and to export oil by 1985.[55] To reach these goals, the military government in 1977 first made peace with the foreign oil companies by ending YPF's monopoly of petroleum-product marketing, which had been in effect since August 1974. Then, to place YPF more directly under government control, the President changed the legal status of YPF from an autonomous state enterprise to a joint-stock company with the state as the only shareholder. Taking the advice of José Martínez de Hoz, his powerful Economics

Minister, Videla named a civilian and longtime critic of the state oil enterprise, Raúl Agustín Ondarts, to head the reorganized company. The YPF bureaucracy and the labor unions opposed this appointment, but the "developmentalists," were clearly in the ascendant in the government. They may not have an easy time operating YPF, however, for they have announced plans to cut its work force from 48,000, the level of April 1977, to 35,000—a cut that may provoke strong opposition from the unions.[56]

Petroleum production increased slightly in 1976 and 1977, but the "developmentalists" in the army view this as only the start of a major oil boom. They aim to invite large-scale foreign petroleum investment back into Argentina to explore for and produce oil both on the mainland and on the continental shelf. According to these plans, various kinds of joint-venture and "risk contract" arrangements will associate YPF with the foreign companies. The oil reserves on the continental shelf are of vast potential: sanguine observers estimate them at up to 31 *billion* cubic meters, or over five times the North Sea reserves. Conditions off the Patagonian shore, however, are extremely difficult. Nonetheless, the military government, according to *The Review of the River Plate*, hopes to transform Argentina

TABLE 6.3. *State and Private Petroleum Production in Argentina, 1958–73*
(Thousands of cubic meters)

| Year | Total domestic production | State fields | | | | Private fields | |
		Government fields	Pct. of total domestic production	Private contractors	Pct. of total domestic production	All private fields	Pct. of total domestic production
1958	5,669	4,964	87.6%	—	—	705	12.4%
1959	7,087	6,127	86.4	327	4.3%	633	9.0
1960	10,153	7,126	70.2	2,465	24.3	562	5.5
1961	13,428	9,134	68.0	3,774	28.1	520	3.9
1962	15,614	10,438	66.9	4,689	30.0	487	3.1
1963	15,445	10,319	66.8	4,704	30.5	422	2.7
1964	15,943	10,779	67.6	4,818	30.2	346	2.2
1965	15,625	10,198	65.3	5,113	33.7	314	2.0
1966	16,656	12,164	73.0	4,198	25.2	294	1.8
1967	18,232	13,772	75.5	4,192	23.0	268	1.5
1968	19,951	15,114	75.7	4,600	23.1	237	1.2
1969	20,681	14,874	71.9	5,595	27.1	212	1.0
1970	22,798	15,379	67.4	7,221	31.7	198	0.9
1971	22,565	16,939	68.9	7,440	30.3	186	0.8
1972	25,193	17,583	69.8	7,429	29.5	181	0.7
1973	24,441	17,450	71.4	6,819	27.9	172	0.7

SOURCES: Instituto Argentino del Petróleo, *El Petróleo en la República Argentina*, p. 5; Villar Araujo, p. 14.

into a member of the elite group of oil-exporting countries. Eight to ten years, and an investment of at least $4 billion, will be required to attain this position.[57] (For a summary of Argentine petroleum production and consumption since 1930, see Tables 6.1, 6.2, and 6.3.)

Political stability will also be necessary, and it is in this area that the military's oil policy is on weak ground. Military rule enjoys little popular support in Argentina; moreover, the army's recent policy of viciously suppressing real and alleged opposition has created tensions and hatreds that spell renewed political instability. Scattered evidence also indicates that opposition exists within the military to the government's oil policy.[58] The generals who head the government despise politicians, but the Peronist movement and the Radical Party both retain a strong popular following. If civilians regain power, Argentine oil policies may well change dramatically again, for economic nationalism enjoys widespread popular support in Argentina, and YPF is the symbol of that nationalism.

Petroleum Nationalism as a Political Ideology

This book has examined the powerful and often decisive impact that nationalism has exerted on Argentine petroleum policy throughout the twentieth century. Petroleum nationalism emerged during the energy crisis of the First World War, when for the first time Argentines began to realize the crucial importance of their undeveloped oil resources. The political elite, tied to a liberal international model, at first regarded the state oil enterprise with disinterest; but after Enrique Mosconi assumed leadership of YPF in 1922, the state firm grew rapidly and became a vertically integrated rival of the foreign oil companies.

Mosconi's role as an ideologist was at least as important to Argentine petroleum history as his successful organization of YPF.[59] Prior to the 1920's, the bulk of Latin American economic thought on development strategies reflected views on the proper functioning of the international economic system that originated in the industrialized countries. The proper role of Latin America, it was assumed, was to produce raw materials for export and to import finished products and capital to finance expansion of the export-oriented economic base. The traditional liberal economists and their allies among the Latin American elites accepted the view that the Latin American petroleum industry would function most rationally as part of an integrated world network of production, refining, and sales—all run by the international oil companies. But Mosconi provided a different interpretation of the world oil industry and a new vision of the role petroleum could play in Latin American economic development. Drawing on ideas that the pioneers of the Argentine oil industry had been developing since 1907, he viewed the multinational oil companies not as bearers of progress and development, but as agents of neocolonial exploitation seeking to despoil the Latin American countries of one of their most essential raw materials and, when necessary, intervening in politics to gain their ends. Only state oil monopolies, he believed, would be able to protect Latin America's petroleum resources in order to promote the industrialization he thought essential to the region's economic well-being and political independence.

Despite his idealism, Mosconi was a highly pragmatic man who combined a faith in the state monopoly idea with a desire for maximum productivity. To achieve this goal, he urged a mixed-ownership plan (limited to Argentine capital) for the YPF monopoly similar to the scheme the British had employed successfully with the Anglo-Persian Oil Company.[60] This aspect of Mosconi's thought never generated political support in Argentina, where advocates of a total state monopoly consistently argued that allowing private investment in the state oil company would open the way for foreign capital to control it. From the early 1920's on, technocratic elements of the military have allied with the YPF bureaucracy and the Radical Party to resist mixed-ownership schemes for fear that private investors would demand reorganization of the enterprise's administration and control over its management.

The portion of Mosconi's ideology that survived to form the core of modern Argentine petroleum nationalism is the state monopoly thesis. During the 1920's, support for this plan was strongest among Argentina's large and underemployed middle class and within the political party that appealed to that class, the Unión Cívica Radical. Yrigoyen gambled the success of his second government on a campaign to grant YPF a total monopoly over Argentine oil production. Although this effort failed—partly because of the impact of the depression and partly because of provincial opposition—the political appeal of a YPF monopoly grew after the 1930 revolution. The government kept petroleum prices low, and as a result consumer support for petroleum nationalism remained strong. By 1935, the foreign producers were confined to their existing holdings, and during the next 40 years proposals to enlarge the foreign petroleum sector aroused formidable political opposition. However logical they may have been in economic terms, the oil policies that Perón and Frondizi launched

20. An example of YPF's nationalistic advertising, 1934: "The Nation Propelling Itself: YPF, The Organ of Power."

in opposition to the state monopoly tradition contributed to both leaders' loss of legitimacy. Petroleum nationalism continues to attract middle-class support, as the Radical Party's enduring commitment to Yrigoyen's principles demonstrates, but during the past quarter century it also has gathered followers within the nation's huge organized labor movement. This popular appeal of YPF has combined with Argentina's high level of mass political mobilization—a legacy of Peronism—to give petroleum nationalism its strength as a political ideology.[61]

The Argentine Model in Latin America

In a broader perspective, Argentine petroleum nationalism has influenced the direction of oil policy in much of the rest of Latin America. Mosconi personally played a major role both in portraying YPF as a model for state-owned oil enterprises in other Latin American countries and in providing them with an ideology of nationalistic oil development. His inter-American trip of 1927–28, during which he attacked the international oil companies and urged a coordinated Latin American petroleum policy, was only the initial step in Mosconi's continental petroleum campaign.

First to follow Argentina's example was Uruguay, a country that depended entirely on imported petroleum and that possessed a strong tradition of economic nationalism and state ownership in such areas as banking, insurance, and utilities dating from the 1910's. Early in 1929, Edmundo Castillo, Uruguay's Minister of Industry, made an official visit to inspect YPF's facilities and to confer with its Director-General. Mosconi gave Castillo a warm welcome and emphasized the economic advantages of a nationally owned refinery; but he also advised his guest that a government corporation should be formed to market the refinery's products. These ideas became the seeds of ANCAP (Administración Nacional de Combustibles, Alcohol y Portland), the state energy corporation that Uruguay's Congress created in 1931. Two years later, the Montevideo government authorized ANCAP to begin constructing both a large refinery and a nationwide retail-outlet system.[62]

YPF and ANCAP cooperated closely during the first years of the Uruguayan state company's operation. The first Uruguayan petroleum engineers, technicians, and chemists received their training as guests of YPF's La Plata refinery, and YPF specialists joined ANCAP to help operate its new refinery during the break-in stages. When the Uruguayan refinery was inaugurated, General Mosconi was among the guests of honor. Despite his illness, Mosconi gave a short, stirring speech with a strong nationalist emphasis. Refineries like ANCAP's, he proclaimed, would form the basis of "the integral independence of Latin America." By 1939, the Uruguayan refinery supplied 90 percent of the republic's gasoline, and to this day ANCAP continues to enjoy a virtual monopoly over refining and marketing in Uruguay.[63]

The example of YPF also made a strong impact in another of Argentina's neighbors, Bolivia. Smarting under its decisive military defeat by Paraguay in the Chaco War (1932–35), Bolivia was the scene of intense nationalist agitation in the late 1930's. The corrupt and ineffective oligarchic governments that had led Bolivia into the Chaco debacle came under severe criticism from young military officers, who seized power in June 1936 and immediately moved to implement a series of social and economic reforms. Nationalization of Standard Oil's holdings was high on their list of priorities. During the war, Standard, which had never produced much oil in Bolivia anyway, transferred much of its equipment to Argentina, refused to refine aviation gasoline, and generally declined to cooperate with the government. The company's attitude, along with the belief of many Bolivians that Standard had promoted the Chaco War in order to build a pipeline to the Paraguay River, brought it under severe criticism. The military government of Colonel David Toro moved swiftly on the oil question. In December 1936 it created Yacimientos Petrolíferos Fiscales Bolivianos (YPFB), a government-owned corporation modeled on the already famous state corporation in Argentina. YPFB's first president, Dionisio Froianini, lauded "the brilliant successes of YPF" as "a highly honorable note not only for the Argentine nation but for all of Latin America." Froianini added that YPF's "norms and principles will guide the Bolivian entity."[64]

Three months after the birth of YPFB, under intense pressure from radical army officers, Toro decreed the expropriation of the Standard Oil Company of Bolivia on the grounds that it had defrauded the government and violated the terms of its concession. Bolivia thus became the first country in Latin America to nationalize a foreign oil company. Although the Bolivians looked to YPF for assistance in organizing the new company's production and marketing its products, the Justo government showed little interest in YPFB. Not until 1940, when the wartime energy crisis loomed, did Buenos Aires begin to extend technical and financial assistance to the Bolivian oil company and permit the duty-free import of YPFB oil. But starved for capital and hindered by transportation obstacles, YPFB developed slowly. Finally, in the 1950's, the Bolivian government again invited foreign companies to invest in petroleum production, although YPFB survived and continued to monopolize both refining and distribution.[65]

In Brazil, which set up a state oil enterprise in 1938 before petroleum was even discovered, Mosconi's ideology and the model of YPF also exercised powerful influence. As in Argentina, the leading figure in the establishment of the Brazilian state oil industry was a military engineer, General Julio Caetano Horta Barbosa. Inspired by Mosconi, Horta Barbosa believed that Brazil's lack of oil production and the foreign companies' control of refining and marketing left his country economically and militarily weak. After becoming Deputy Chief of Staff in 1937, Horta urged President Getúlio Vargas to make the oil industry a state monopoly, starting with refining. Mosconi's insistence that a state oil company must supply a large share of the market in order to control prices successfully made a particularly strong impression on the Brazilian general, who drew up a plan to nationalize production and refining and to leave only importing and distribution to foreign capital. Although Vargas at first was skeptical, the military hierarchy backed the plan and on May 1, 1938, the President signed a decree nationalizing production and placing refining under control of a newly created Conselho Nacional do Petróleo (CNP). General Horta became the CNP's first president. Private refineries continued to operate, but the CNP, stimulated by the discovery of oil in Bahia in 1939, began planning to establish its own refinery.[66]

Mosconi enjoyed greater official prestige in the Brazil of Vargas than in the Argentina of Justo. On July 14, 1938, the Academy of Sciences and the Arts of Rio de Janeiro recognized Mosconi's work by according him its gold medal. Because Mosconi was too ill to travel to Rio to receive it, the Brazilian Consul General made the presentation at Mosconi's home in Buenos Aires.[67] The following year, General Horta and a group of aides traveled to Argentina to study YPF's operations. In addition to visiting the Comodoro Rivadavia oil fields, Horta had a talk with Mosconi, who again pressed home his point that state refineries would be essential to enable CNP to set prices in the Brazilian market. The Second World War and domestic political opposition intervened to frustrate the plans of Horta, who resigned in 1943, with the result that the first Brazilian state refinery did not begin production until 1951. Meanwhile, General Horta, joined by both civilian and military nationalists, was waging a determined battle to expand the powers of CNP. During a 1947 debate at Rio's Military Club, Horta argued that the Argentine and Mexican experiences proved that state oil monopolies benefited the entire national economy whereas private monopolies, such as that tolerated by Venezuela, left countries shackled to imperialism.[68]

Ultimately, Horta's ideas triumphed, when during the final presidency of Getúlio Vargas the Brazilian Congress voted in 1953 to create Petrobrás, Brazil's state oil monopoly. Although the existing private refineries continued to operate, Petrobrás henceforth monopolized all other stages of the national petroleum industry except distribution. Despite troublesome capital shortages, Petrobrás grew rapidly and became one of the world's largest enterprises. By 1966, it refined 93 percent of Brazil's consumption.[69] Nonetheless, Brazil remained reliant on imports for at least half of its crude petroleum, which meant that its oil import bill rose to $2 billion after the world oil crisis struck in 1973. Brazil's oil bill was $3.5 billion in 1974—close to a third of the country's export earnings. Although free-market advocates in Brazil's military government have begun to suggest curtailment of the Petrobrás monopoly in order to permit foreign companies to search for oil, particularly offshore, such suggestions touch off powerful opposition—for like YPF, Petrobrás has become a symbol of nationalism. But in 1976 the regime in Brasília decided to deal with the foreign companies.[70]

The model of YPF was strongest in Uruguay, Bolivia, and Brazil, but it also influenced petroleum policy in Mexico. In 1930, two years after Mosconi's visit to Mexico City, the Ministry of Industry's *Boletín del Petróleo* published an editorial portraying YPF in highly favorable terms, lauding the work of Mosconi, and arguing that Mexico should follow the Argentine example. Although Mexican petroleum nationalism had deep roots dating from the 1910–17 revolutionary period, this article marked the first occasion that the Mexican government publicly discussed the idea of establishing a state oil monopoly.[71] The state company idea gathered force and culminated in the famous expropriation decrees of March 18, 1938, when President Lázaro Cárdenas ordered the immediate nationalization of the entire Mexican oil industry under the state petroleum monopoly, PEMEX. Cárdenas, who had heard Mosconi deliver his famous "two-rope speech" at the National University in 1928, espoused an ideology of petroleum nationalism similar to that of the Argentine general.[72]

By the mid-1960's, 11 Latin American countries had formed state oil

companies that played increasingly important roles in their national economies. They produced 18 percent of Latin America's crude output, owned
two-thirds of the region's total refining capacity, and controlled 53 percent
of petroleum-product distribution. [73] But it was not until Venezuela na-
tionalized its entire oil industry in 1975 that the Latin American state
companies controlled a clear majority of the region's oil production.
Whether Venezuela will be willing or able to undertake investments in the
energy sector elsewhere in Latin America remains unclear, although Venezuelan support for a new cooperative regional energy association, *Organización Latinoamericana de Energía* (OLADE), indicates that Caracas may desire
a continental petroleum policy. [74] What the Venezuelan nationalization
clearly shows is that petroleum nationalism remains powerful as Latin
America enters the final decades of the twentieth century.

Reference Matter

Notes

Complete authors' names, titles, and publishing data for sources cited in the Notes are given in the Bibliography, pp. 217–33. The citations "Argentina," "Great Britain," and "United States" followed by bracketed numbers indicate official or semiofficial publications listed under those countries in the Bibliography.

Notes to Chapter 1

1. The pattern of concentrated land ownership in the key province of Buenos Aires was typical of the pampas provinces. In 1928, only 1,041 estates accounted for a third of the province's area, including much of its best grazing and farming land. See Oddone, pp. 182–85. Before the First World War, about 70 percent of the farmers in the pampas provinces were tenants. On this point see Pavlovsky, p. 27. An excellent source of economic data for the 1870–1918 period is Ernesto Tornquist & Cía., *El desarrollo económico de la República Argentina en los últimos cincuenta años*, published in 1920. For the agricultural boom, see pp. 2–9 and 20.

2. McCrea et al., p. 59; Tornquist, p. 134.

3. Vernon L. Phelps, p. 108; Duvall, pp. 284–85.

4. Bunge, *Ferrocarriles argentinos*, p. 435; L. Brewster Smith et al., pp. 74–75; Vernon L. Phelps, p. 108.

5. Bunge, *Ferrocarriles argentinos*, pp. 126–27.

6. For a careful analysis of foreign investment, see Vernon L. Phelps, pp. 108–9, and Ford, "British Investment and Argentine Economic Development." A 1918 accounting computed total foreign investments (including mortgages) of 3.882 billion gold pesos. See Martínez, p. 1399.

7. Hobsbawm, p. 121. For an extensive analysis of British influence in Argentina, see Fodor and O'Connell.

8. Solberg, "The Tariff and Politics in Argentina," p. 261.

9. Díaz Alejandro, p. 514.

10. Maizels, p. 533.

11. For a masterful analysis of the socioeconomic underdevelopment of the northwest, see Rutledge. Also see Chavarría and Bazán.

12. Fraser, p. 26. For a careful analysis of price levels, see Scobie, pp. 141–

42. An example of a historian who has overlooked price levels in celebrating Argentina's pre-1914 economic growth is Ferns, p. 17.

13. Bunge, *Una nueva Argentina*, p. 361. Also see Juan B. González; and Scobie, pp. 156–58.

14. Brady, *Railways of South America*, pp. 24–25. Hydroelectric power, potentially a rich source of energy, remained undeveloped in Argentina. Bain and Read, p. 182.

15. Platt, p. 246; Sargent, p. 28.

16. Bradley, p. 48; Fodor and O'Connell, p. 16.

17. Hermitte, "Carbón, petróleo y agua," p. 121. For early coal prospecting in Argentina, see *The Review of the River Plate*, 57 (Jan. 26, 1922), p. 217; and L. Brewster Smith et al., pp. 60–61.

18. Hermitte, *Las investigaciones*, pp. 33–34.

19. Gibb and Knowlton, p. 198. Also see Kaplan, *Petróleo*, p. 47; and Brady, *Argentine Petroleum Industry*, pp. 1, 17.

20. Kaplan, "La primera fase," pp. 789–91.

21. Méndez, pp. 105–6; Kaplan, *Petróleo*, pp. 54–55; *The Review of the River Plate*, 51 (Apr. 25, 1919), p. 1005.

22. Argentina, [33], p. 338.

23. Davel, p. 151; Hileman, "Los yacimientos petrolíferos de Cacheuta," p. 71; Hermitte, "Carbón, petróleo y agua," pp. 128–30.

24. For the town's history, see Mármora, pp. 27–28. For its appearance in 1907, see Martini, p. 39.

25. See Miatello's excellent study of the Comodoro Rivadavia region, especially pp. 32–33, and 70.

26. *Ibid.*, p. 54; Hermitte, "Carbón, petróleo y agua," pp. 74, 131.

27. For Hermitte's 1904 disclosure, see Hermitte, "Carbón, petróleo y agua," p. 74. For the 1903 incident, see Craviotto, p. 524. For a strong statement of the argument that the government was in fact seeking oil, see Krause and Krause, pp. 8–14, 34–35. Alsacio Cortázar suggests (pp. 32–33) that the British, desiring to secure naval fuel supplies, encouraged the Argentine government to search for oil in Patagonia. Also see Martini, p. 19.

28. Martini, p. 114.

29. The 1903 law gave the government authority to prohibit "sale of lands containing known deposits of salt, coal, petroleum, or mineral waters." See Craviotto, p. 567. For the text of Figueroa Alcorta's decree, see the valuable collection of documents in Argentina, [34], p. 18

30. Argentina, [11], p. 7; Argentina, [34], pp. 25–26; Kaplan, *Petróleo*, p. 41.

31. Melo, pp. 102–3; Galletti, p. 43.

32. For a penetrating analysis of the relations between the Radical Party and the middle classes, see Rock, "Radical Populism," pp. 66–70, and Rock, *Politics in Argentina*, pp. 19–20. The Radical Party's early years receive close attention in Gallo and Sigal, especially pp. 135–37, 163–65.

33. For Yrigoyen's early leadership of the party, see Caballero, especially pp.

38–39; del Mazo, vol. 1, pp. 57–113; Luna, *Yrigoyen*, pp. 117–22; and Rock, *Politics in Argentina*, pp. 50–55.

34. Melo, pp. 102–3.

35. Cárcano, pp. 147–55.

36. Yrigoyen made the famous remark: "Si el gobierno nos da garantías, concurriremos a las urnas." (Quoted in Cárcano, p. 156.) Also see del Mazo, vol. 1, p. 133.

37. Argentina, [2], Sesiones ordinarias, vol. 2 (Sep. 2, 1908), pp. 993–94.

38. For Figueroa Alcorta's message to Congress, see Argentina, [5], Sesiones ordinarias, vol. 1 (Sep. 2, 1909), p. 542.

39. The debates are in Argentina, [5], Sesiones ordinarias, vol. 1 (Sep. 4, 1909), pp. 570–80; and Argentina, [2], Sesiones ordinarias, vol. 2 (Aug. 29, 1910), pp. 244–47.

40. Craviotto, p. 544.

41. Velarde, pp. 193–95; London *Times*, Aug. 17, 1925, Argentine Supplement, p. xv. For a list of the oil lands the government conceded to private ownership, see Argentina, [2], Sesiones ordinarias, vol. 3 (Aug. 17, 1927), p. 821.

42. Argentina, [7]. For summary treatments of the mining code, see Brady, *Argentine Petroleum Industry*, p. 19; and Velarde, pp. 22–23, 50–54.

43. "The Argentine Oil Industry," *Comments on Argentine Trade*, 4 (Mar. 1925), p. 25; Velarde, pp. 193–95. Velarde, a professional petroleum engineer and Bureau of Mines employee, put his considerable field experience to good effect in his careful examination of how the existing laws operated as of 1922.

44. Velarde, pp. 58–60.

45. Méndez, p. 177.

46. Sáenz Peña, vol. 2, pp. 211–12; Hollander, pp. 255–59, 271.

47. For the text of Sáenz Peña's decree, see Argentina, [26], pp. 9–13.

48. Argentina, [10], pp. 10, 17, 40; Argentina, [11], pp. 5, 29.

49. For Sáenz Peña's message to Congress of Sep. 26, 1911, see Sáenz Peña, vol. 2, p. 208.

50. Argentina, [9], p. 32; Argentina, [34], p. 13.

51. Repetto, p. 158; Argentina, [9], pp. 38–40.

52. Argentina, [11], pp. 27–38; London *Times*, Engineering Supplement, Dec. 17, 1913, p. 28; London *Times*, Trade Supplement, Dec. 13, 1919, p. 332.

53. For the 1912 budget, see Argentina, [11], pp. 24–26. Also see Argentina, [9], p. 34.

54. Sáenz Peña, vol. 2, p. 210.

55. Argentina, [2], Sesiones ordinarias, vol. 2 (Jul. 30, 1913) pp. 604–5.

56. For analysis of Huergo's career and of the Buenos Aires port dispute, see Scobie, pp. 72–89.

57. Quoted in Casal, p. 20. Also see Kaplan, "La primera fase," pp. 785–88.

58. Cáceres Cano, p. 37. For the text of the decree, see Argentina, [26], pp. 20–21. Also see Craviotto, pp. 546, 557. Private investors sharply condemned this decree. For the views of Carlos A. Tornquist, see *The Review of the River Plate*, 60 (Aug. 24, 1923), p. 443.

59. Argentina, [2], Sesiones ordinarias, vol. 3 (Aug. 17, 1927), p. 821; Kaplan, "La primera fase," p. 792; Rumbo, pp. 22–24. For de la Plaza's message asking Congress to turn over the operation and administration of the state's reserve to private interests, see Argentina, [5], Sesiones ordinarias, vol. 1 (Jul. 16, 1914) pp. 145–47.

60. Demarchi's plan is in Argentina, [2], Sesiones ordinarias, vol. 1 (Jun. 15, 1914), pp. 777–80. Also see Argentina, [2], Sesiones ordinarias, vol. 2 (Aug. 22, 1913), pp. 1034–37.

61. Argentina, [9], pp. 55–56.

62. Argentina, [34], p. 43.

63. London Times, Trade Supplement, Dec. 13, 1919, p. 332; Méndez, pp. 131–32; Argentina, [34], pp. 384–87.

Notes to Chapter 2

1. For a thorough analysis of the economic impact of the European crisis on Argentina, see Ford, *The Gold Standard*, pp. 170–88. Also see Salera, pp. 21–23; and Vernon L. Phelps, pp. 24–26, 36–37.

2. Tulchin, "The Argentine Economy," pp. 901–3; also see L. Brewster Smith et al., and Bianco.

3. The memoirs of labor leader Cipriano Reyes provide a graphic description of the impact of the wartime economic conditions on the masses (pp. 15–27). For the socioeconomic situation in the interior during this period, see Chavarría, p. 165; and Olguín, pp. 209–10.

4. Tornquist, p. 183; Ferrer, p. 80; Peter H. Smith, *Politics and Beef*, p. 69.

5. L. Brewster Smith et al., pp. 85–95; Tornquist, pp. 148–60.

6. Great Britain, [5], p. 12. For further analysis of industry during the First World War, see Di Tella and Zymelman, pp. 306–12; Cartavio; and Dorfman, pp. 135–36.

7. Havens, p. 649; *The Review of the River Plate*, 47 (Feb. 23, 1917), p. 407.

8. Méndez, pp. 79–81; *The Review of the River Plate*, 50 (Aug. 9, 1918), p. 337.

9. The Buenos Aires–Western Railway reported that its fuel bill rose from £193,909 in 1913–14 to £739,561 in 1918–19. See *La Prensa*, Jan. 5, 1920, p. 6. For additional analysis of the impact of the fuel crisis on the railways, see *The Review of the River Plate*, 48 (Nov. 23, 1917), p. 1259; and Brady, *Railways of South America*, p. 24.

10. Bunge, *Ferrocarriles argentinos*, pp. 201–4; Tornquist, p. 94.

11. Bunge, *Ferrocarriles argentinos*, p. 199; Vernon L. Phelps, p. 16.

12. Dreier, p. 12; Havens, p. 650.

13. Jane, pp. 422–32.

14. Dorfman, p. 142.

15. Yrigoyen's decree authorizing the exchange is in Argentina, [26], pp. 69–70.

16. Argentina, [30], p. 21; Argentina, [31], p. 26.

17. "Argentine Minerals, Opinions of Engineer Hermitte," *The Review of the River Plate*, 47 (Apr. 20, 1917), p. 893.

18. Bradley, p. 40.

19. *La Prensa*, Nov. 18, 1917, p. 8; Argentina, [30], p. 20; Méndez, p. 59.

20. Argentina, [30], p. 59.

21. An exception to the general intellectual commitment to private enterprise during this period concerned the British-owned railways. Some writers argued in favor of nationalization and state ownership as early as 1918. See Wright, p. 199. Representative intellectual precursors of economic nationalism during the First World War included Ruiz Guiñazú, p. 51; Bianco, pp. 44–46, 245–68; Quesada, pp. 72–73. Also see Tulchin, "The Argentine Economy," p. 46.

22. Ugarte, p. 8.

23. Johnson, "The Ideology," pp. 131–32. Alejandro Bunge acknowledged his intellectual debt to List in *Una nueva Argentina*, p. 238.

24. Bunge, a remarkable scholar, published at least 389 articles and a dozen books between 1914 and 1940. See the bibliography in Bunge, *Una nueva Argentina*, pp. 487–513. A professor in the Faculty of Economics at the University of Buenos Aires, Bunge in 1918 founded the prestigious *Revista de Economía Argentina*, which he edited until his death in 1943. As Director-General of the Bureau of National Statistics from 1916 to 1922, he introduced the use of index numbers into Argentina and began to keep the country's first methodical statistical series on the cost of living and on real wages. See United States, [3], 835.501/2, Jan. 17, 1922, Robertson (Buenos Aires) to Secretary of State; Bunge, *Una nueva Argentina*, pp. 15–19; and Imaz, "Alejandro E. Bunge," pp. 547–48.

25. For a summary of Bunge's economic ideas during the First World War, see Moyano Llerena, p. 39; Imaz, "Alejandro E. Bunge," pp. 554–55; and Bunge, *Una nueva Argentina*, pp. 230–38.

26. Solberg, "The Tariff," pp. 271–74, 280–82.

27. For an overview of the emergence of economic nationalism during the 1930's, see Falcoff, "Intellectual Currents," pp. 126–28. Bunge emphasized his commitment to private capitalism in Bunge and Sasot, pp. 53–64.

28. Oneto published a collection of his articles, q.v. For evaluations of his contributions by later petroleum nationalists, see Silenzi de Stagni, *El petróleo argentino*, p. 49; and Casal, p. 23.

29. Kaplan, "La primera fase," pp. 805–6; Argentina, [2], Sesiones ordinarias, vol. 2 (Aug. 18, 1916), pp. 1410–13.

30. Rock, *Politics in Argentina*, p. 97; Cantón, vol. 1, pp. 37–41.

31. Yrigoyen's use of the intervention policy to further the political ends of his party receives detailed examination in Robert Richard Smith, "Radicalism in the Province of San Juan."

32. A complete list of Yrigoyen's interventions appears in Bucich Escobar, p. 437.

33. For an interpretation of Yrigoyen as *caudillo*, see Sánchez Viamonte. For Yrigoyen's political philosophy, see Yrigoyen, *Mi vida y mi doctrina*, pp. 38,

46–47, 56–57. An excellent introduction to *krausismo* is Pike, pp. 310–11. For analyses of Yrigoyen's use of *krausismo*, see Sommi, pp. 292–95; and Landa, pp. 254–57.

34. *La Prensa*, Apr. 9, 1919, p. 11; Apr. 12, 1919, p. 9.

35. *La Prensa*, Feb. 15, 1919, p. 5; United States, [3], 835.00/266, Dec. 16, 1921, White (Buenos Aires) to Secretary of State.

36. Fennell, p. 116. For a thorough analysis of Yrigoyen's control of the Radical political machine, see Rock, "Machine Politics," pp. 233–56.

37. U.S. Consul White reported in 1921 that during serious labor disturbances that shut down the port of Buenos Aires Yrigoyen was preoccupied with an intervention in the province of Buenos Aires. When one of White's friends asked Yrigoyen whether he was concerned with the port situation, he responded, "No! But I am very concerned about Crotto!" [Governor of Buenos Aires]. United States, [3], 835.00/241, Aug. 31, 1921, White (Buenos Aires) to Secretary of State.

38. Frank, *America Hispana*, p. 123.

39. For Yrigoyen's family background, see Luna, *Yrigoyen*, pp. 15–16. His landholdings and economic interests are considered in *ibid.*, pp. 57–59; and in Sommi, pp. 292–95.

40. Merkx, p. 136.

41. Peter H. Smith, *Politics and Beef*, pp. 49, 130–31.

42. See *The Review of the River Plate*, 49 (Jan. 18, 1918), p. 147.

43. For expressions of farmers' opinions, see Holm, p. 35; and Argentina, [2], Sesiones extraordinarias, vol. 7 (Nov. 26, 1919), p. 447.

44. Solberg, "Rural Unrest," pp. 18–52; Solberg, "The Tariff," pp. 260–84.

45. United States, [3], 835.00/266, Dec. 16, 1921, White (Buenos Aires) to Secretary of State.

46. Buchanan, pp. 42–46. For finance policy, see Spinelli, pp. 6–7; and Argentina, [4], vol. 1, p. 200.

47. *La Epoca*, Oct. 30, 1916, p. 1. For Salaberry's proposal, see Argentina, [2], Sesiones extraordinarias, vol. 4 (Dec. 11, 1916), p. 2791.

48. Argentina, [2], Sesiones extraordinarias, vol. 5, (Jan. 26, 1917), pp. 4693–4700, 4952–75; vol. 5 (Feb. 15, 1917), p. 5138; vol. 5 (Feb. 16, 1917), p. 5191. *The Review of the River Plate*, 47 (Feb. 23, 1917), p. 407, and 47 (Mar. 9, 1917), p. 519; Tulchin, "The Argentine Economy," p. 967.

49. Pegazzano, pp. 321–22.

50. *La Protesta*, Oct. 5, 1917, p. 3.

51. "The Argentine Oil Industry," *Comments on Argentine Trade*, 4 (Mar. 1925), p. 9; Gómez, pp. 12, 30.

52. Hanighen, pp. 166, 171; Miatello, p. 54.

53. Great Britain, [1], 371/8418, Jan. 16, 1923, Wilson (Buenos Aires) to Foreign Office, pp. 19–20; Hanighen, p. 168; Miatello, p. 54.

54. *La Vanguardia*, Nov. 1, 1917, p. 1; *La Prensa*, Nov. 1, 1917, p. 8; Argentina, [14], p. 48; Argentina, [15], pp. 54–55.

55. *La Vanguardia*, Nov. 1, 1917; Argentina, [14], p. 55; *La Razón*, Dec. 11, 1919 (clipping in Biblioteca Tornquist, Buenos Aires).

56. Miatello, pp. 32–33, 70; Argentina, [14], p. 55; Great Britain, [1], 371/9504, Apr. 4, 1924, Wilson (Buenos Aires) to Foreign Office, p. 9; London *Times*, Aug. 17, 1925, Argentine Supplement, p. xv. Yrigoyen's decree expropriating Behr's Estancia is in Argentina, [26], pp. 94–95.

57. *La Prensa*, Nov. 1, 1917, p. 8; "Las condiciones obreras en Comodoro Rivadavia," *El Petróleo Argentino: Revista Quincenal de la Industria Nacional del Petróleo y sus Derivados*, 1 (Jun. 10, 1923), pp. 11–12; Miatello, p. 70.

58. *La Prensa*, Nov. 1, 1917, p. 8; Argentina, [14], p. 82.

59. *La Prensa*, Nov. 1, 1917, p. 8; "De Comodoro Rivadavia," *El Petróleo Argentino: Revista Quincenal de la Industria Nacional del Petróleo y sus Derivados*, 1 (Aug. 10, 1923), p. 17; Hanighen, p. 166; *La Vanguardia*, Oct. 23, 1917, p. 2.

60. *La Vanguardia*, Nov. 1, 1917, p. 1; Gómez, p. 29.

61. *The Review of the River Plate*, 48 (Nov. 2, 1917), p. 1071; *La Protesta*, Feb. 5, 1920, p. 3.

62. *La Prensa*, Nov. 1, 1917, p. 8; *La Vanguardia*, Oct. 23, 1917, p. 2; *La Protesta*, Nov. 2, 1917, p. 1, and Nov. 9, 1917, p. 1; *The Review of the River Plate*, 48 (Oct. 26, 1917), p. 1037; United States [3], 835.5045/36, Nov. 8, 1917.

63. *La Prensa*, Nov. 3, 1917, p. 5; Nov. 4, 1917, p. 8; Nov. 5, 1917, p. 8; Nov. 23, 1917, p. 5. *La Protesta*, Nov. 2, 1917, p. 1.

64. *La Prensa*, Nov. 1, 1917, p. 8; Nov. 3, 1917, p. 8; Nov. 8, 1917, p. 9; Nov. 9, 1917, p. 9. *La Vanguardia*, Nov. 8, 1917, p. 1. *The Review of the River Plate*, 48 (Nov. 9, 1917), p. 1153.

65. *La Prensa*, Nov. 9, 1917, p. 11; Nov. 10, 1917, p. 10; Nov. 15, 1917, p. 8. *La Vanguardia*, Nov. 10, 1917, p. 2; Nov. 13, 1917, p. 3. Buchanan, p. 51.

66. *La Prensa*, Nov. 14, 1917, p. 9; Nov. 15, 1917, p. 8. *La Vanguardia*, Nov. 13, 1917, p. 3.

67. Argentina, [31], p. 5; Buchanan, pp. 55–56.

68. For analysis of this administrative reorganization, see the report of congressional investigators in Argentina, [4], vol. 1, pp. 641–45. For Captain Fliess's opinions of the results of the reorganization, see Fliess, pp. 15–16. Also see United States, [3], 835.002/38, Mar. 19, 1917, Stimson (Buenos Aires) to Secretary of State. Paul B. Goodwin has suggested that in his railway policy Yrigoyen similarly fostered administrative confusion or at least tolerated it in order to retain ultimate power in his own hands. See Goodwin, "The Politics of Rate-Making," p. 258.

69. Fliess, p. 18; *La Prensa*, Dec. 21, 1919, p. 9; *The Review of the River Plate*, 52 (Oct. 31, 1919), p. 1131; Argentina, [13], pp. 3, 27; Argentina, [15], p. 31.

70. *The Review of the River Plate*, 50 (Aug. 30, 1918), p. 562; Argentina, [14], p. 4.

71. For a vivid description of the difficulties of work at Plaza Huincul, see Argentina, [12], pp. 8–9.

72. *The Review of the River Plate*, 50 (Oct. 4, 1918), p. 875; 51 (May 30, 1919), p. 1331. Brady, *Argentine Petroleum Industry*, p. 10.

73. *The Review of the River Plate*, 50 (Nov. 29, 1918), p. 1394; *La Prensa*, Dec. 9, 1918, p. 2, and Dec. 17, 1918, p. 11.

74. *La Prensa*, Dec. 18, 1918, p. 7; Dec. 24, 1918, p. 10; Dec. 30, 1918,

p. 9. *The Review of the River Plate*, 50 (Dec. 20, 1918), p. 1553; 50 (Dec. 27, 1918), p. 1615; 51 (Jan. 10, 1919), p. 95.

75. For the text of De Veyga's proposal, see Argentina, [2], Sesiones ordinarias, vol. 1 (Jun. 26, 1916), pp. 586–92.

76. For Uriburu's address, see Argentina, [2], Sesiones ordinarias, vol. 5 (Sep. 18, 1917), pp. 267–68. Also see Argentina, [2], Sesiones ordinarias, vol. 3 (Aug. 1, 1917), p. 45; and vol. 4 (Aug. 21, 1917), pp. 29–33.

77. Kaplan, *Petróleo*, p. 50; Méndez, p. 103; Argentina, [2], Sesiones ordinarias, vol. 1 (Jul. 17, 1916), pp. 652–53.

78. By 1917, considerable evidence existed that Standard and other oil companies were intervening financially in the Mexican Revolution to support the enemies of nationalist President Venustiano Carranza. At Tampico, the oil companies paid $15,000 a month to local boss General Manuel Peláez, an opponent of Carranza who did not enforce the president's oil taxes. Though Carranza charged that the companies were deliberately encouraging counterrevolution, the State Department did not oppose the payoffs. See Meyer, pp. 99–101; and Robert Freeman Smith, pp. 101–4.

79. Argentina, [2], Sesiones extraordinarias, vol. 5 (Jan. 19, 1917), pp. 4382–86, 4394, 4401.

80. *The Review of the River Plate*, 47 (Jan. 19, 1917), p. 155.

81. Compañía Nacional de Petróleos, p. 8; Argentina, [2], Sesiones extraordinarias, vol. 5, (Feb. 14, 1917), p. 4408–16; Argentina, [5], Sesiones extraordinarias, vol. 2 (Feb. 14, 1917), p. 564.

82. Argentina, [3], pp. 57–59, 72–73; Great Britain, [5], pp. 17–18.

83. Argentina, [2], Sesiones extraordinarias, vol. 5 (Jan. 22, 1917), p. 4415; Argentina, [3], p. 55; Méndez, p. 117.

84. Argentina, [3], p. 73.

Notes to Chapter 3

1. De Novo; Feis, p. 3.

2. Sir Edward's article, published in the September 1919 issue of *Sperling's Journal*, is quoted in Davenport and Cooke, p. ix. For additional analyses of the postwar oil scare, see Tugendhat, pp. 74–75; and O'Connor, pp. 220–22.

3. London *Times*, April 29, 1921, p. 11; Great Britain, [4], p. 2.

4. Deterding, pp. 83–84; Fanning, p. 5; Lieuwen, pp. 18–19; Great Britain, [3], pp. 10–13.

5. Wilkins, "Multinational Oil Companies," p. 428; De Novo, p. 869; Fanning, p. 3; Pinelo, p. 21.

6. As early as 1922 it was stated in the London *Times* that "peace between the oil interests in the United States and the British Empire has now been concluded" (London *Times*, Jan. 9, 1922, p. 10). Also see Hogan, pp. 195–96; Tulchin, *The Aftermath of War*, pp. 118–33; and Feis, pp. 9–10. Despite the State Department's general policy of entente with the British, important exceptions appeared

in some countries, e.g., Colombia. Professor Stephen J. Randall believes that "the record is unequivocal that the State Department worked determinedly to keep British interests from gaining concessions on state lands in Colombia during the 1920's and 1930's" (letter to the author, Oct. 14, 1976). Also see Randall, p. 185.

7. Fanning, p. 12; Wilkins, "Multinational Oil Companies," p. 425.

8. Penrose, p. 93.

9. Fanning, p. 12; *The Petroleum Almanac*, pp. 295, 298; "Resumen estadístico de la economía argentina," *Revista de Economía Argentina*, 20 (Nov. 1938), p. 323.

10. Pinelo, pp. 24–25, 34; Wilkins, "Multinational Oil Companies," pp. 438–39.

11. Klein, pp. 49–53; Gibb and Knowlton, pp. 382–83, 646; Braden, pp. 72–73. Also see *The Review of the River Plate*, 62 (Oct. 17, 1924), p. 1035.

12. Ownership of the Colombian firms furnishes an excellent example of the complex legal arrangements U.S. multinationals employed in Latin American petroleum. The Tropical Oil Co. was operated by the International Petroleum Company (IPC), Canada, which in turn was a subsidiary of Imperial Oil and of Jersey Standard. The pipeline company, the Andean National Corporation, was also controlled by IPC. See Randall, pp. 180–81. Also see Wilkins, "Multinational Oil Companies," pp. 430–31, 441–42; and Rippy, pp. 20–35.

13. Lieuwen, pp. 39–44. The quote is from p. 29. Also see Hartshorn, pp. 270–71; Denny, p. 88; and Wilkins, "Multinational Oil Companies," p. 435.

14. *The Petroleum Almanac*, p. 305.

15. Lieuwen, pp. 39–40; Hartshorn, p. 271; Tugwell, p. 38.

16. Brady, *Argentine Petroleum Industry*, p. 1.

17. Craviotto, p. 549.

18. American Petroleum Institute, *Petroleum Facts and Figures*, pp. 20–21, 27. For background on Jersey Standard's decision to move into Argentine exploration and production, see Gibb and Knowlton, p. 381, Wilkins, "Multinational Oil Companies," pp. 433–34; and Sampson, pp. 33–34.

19. For the text of the proposed legislation, see Argentina, [2], Sesiones ordinarias, vol. 5 (Sep. 27, 1919), pp. 664–66. For background on the 1920 congressional elections, see Walter, *The Socialist Party*, pp. 162–63, 166.

20. Argentina, [2], Sesiones ordinarias, vol. 5 (Sep. 27, 1919), p. 664.

21. Rumbo, p. 68.

22. Historians sympathetic to the Radical Party have interpreted the 1919 proposals as strongly nationalistic in scope and intent. See del Mazo, vol. 1, p. 185; and Frondizi, *Petróleo y política*, pp. 94–97.

23. Buchanan, pp. 91–97.

24. Craviotto, p. 550. Law 10,273 of 1917 abolished stipulations in the mining code that required owners of concessions to employ a minimum number of workers. Instead, they were required to invest 40,000 pesos during the first four years of the concession and to pay an annual tax of 100 pesos per concession.

25. United States, [3], 835.6363/80, Sep. 29, 1921, Sinclair Exploration Co.

to Secretary of State; 835.6363/100, Dec. 29, 1921, Eberly (Standard Oil Co. of New Jersey, Buenos Aires) to Secretary of State. *The Review of the River Plate*, 59 (May 18, 1922), p. 1218.

26. United States, [3], 835.6363/101, Dec. 23, 1921, White (Buenos Aires) to Secretary of State; Buchanan, p. 133.

27. Brady, *Argentine Petroleum Industry*, p. 7; Wilkins, "Multinational Oil Companies," p. 433; Buenos Aires *Standard*, Jan. 4, 1924 (clipping in Biblioteca Tornquist, Buenos Aires). According to Hollander, p. 334, the German investors served as a front for Jersey Standard, which indirectly controlled Astra.

28. *The Review of the River Plate*, 53 (Jun. 25, 1920), p. 1683; Brady, *Argentine Petroleum Industry*, p. 8.

29. London *Times*, Trade Supplement, Jul. 24, 1920, p. 505; Goodwin, *Los ferrocarriles británicos*, p. 179.

30. The terms of the proposal are in Great Britain, [1], 371/4410, May 14, 1920, Macleay (Buenos Aires) to Foreign Office. Also see United States, [3], 835.6363/16, Jan. 9, 1920, Bursley (London) to Secretary of State.

31. London *Times*, Trade Supplement, Dec. 13, 1919, p. 332.

32. Great Britain, [1], 371/4410, May 14, 1920, Macleay (Buenos Aires) to Foreign Office; 371/5521, Mar. 7, 1921, Macleay (Buenos Aires) to Foreign Office.

33. *Ibid.*, 371/4410, May 14, 1920, Macleay (Buenos Aires) to Foreign Office. Also see *The Review of the River Plate*, 54 (Jul. 9, 1920), pp. 89, 91.

34. *The Review of the River Plate*, 57 (Jan. 20, 1922), p. 171; United States, [3], 835.6363/62, Apr. 4, 1921, Stimson (Buenos Aires) to Secretary of State; Brady, *Argentine Petroleum Industry*, p. 81. Great Britain, [1], 371/9504, Apr. 4, 1924, Wilson (Buenos Aires) to Foreign Office, p. 13.

35. United States, [3], 835.6363/29, Jul. 30, 1920, Wiley (Buenos Aires) to Secretary of State. A State Department investigation of Gage's background revealed a checkered career as representative of various U.S. business interests in Latin America. See 835.6363/60, Mar. 4, 1921, Bannerman (New York) to Sharp (Chief Special Agent, Department of State). The Colombian lands passed into the control of Jersey Standard's affiliate, the Tropical Oil Co. See Braden, pp. 72–73.

36. Klein, pp. 49–52. The quote is on p. 51. Also see Gibb and Knowlton, pp. 382–83; and United States, [3], 835.6363/28, Aug. 26, 1920, Wadsworth (Buenos Aires) to Secretary of State. Braden discloses that he worked together with Gage and Philadelphia banker George Kendrick in the Bolivian land acquisitions. After Kendrick experienced financial difficulties, Braden sold the properties to Jersey Standard. Braden, p. 73.

37. United States, [3], 835.6363/28, Sep. 3, 1920, Colby (Washington) to Braden (New York); 835.6363/31, Sep. 9, 1920, Braden (New York) to Secretary of State.

38. *Ibid.*, 835.6363/29, Jul. 29, 1920, Gage (Buenos Aires) to Wiley (Buenos Aires); 835.6363/33, Sep. 29, 1920, Wiley (Buenos Aires) to Secretary of State; 835.6363/41, Dec. 13, 1920, Stimson (Buenos Aires) to Secretary of State.

39. *Ibid.*, 835.6363/29, Jul. 29, 1920, Gage (Buenos Aires) to Wiley (Buenos Aires); 835.6363/20, Jun. 8, 1920, Stimson (Buenos Aires) to Secretary of State;

835.6363/28, Wadsworth (Buenos Aires) to Secretary of State; 835.6363/21, Jun. 8, 1920, Report on the Bolivia-Argentine Exploration Company by Bureau of Foreign and Domestic Commerce.

40. *Ibid.*, 835.6363/28, Sep. 10, 1920, Colby (Washington) to U.S. Embassy (Buenos Aires).

41. *Ibid.*, 835.6363/55, Mar. 7, 1921, Office of the Foreign Trade Adviser, Department of State, to Secretary of State; 835.6363/45, Feb. 1, 1921, Memo of phone conversation with Spruille Braden; 835.6363/46, Feb. 4, 1921, Memo of phone conversation with George Kendrick.

42. *Ibid.*, 835.6363/40, Jan. 7, 1921, Braden (New York) to Secretary of State; 835.6363/44, Jan. 28, 1921, Stimson (Buenos Aires) to Secretary of State; 835.6363/46, Feb. 4, 1921, Memo of phone conversation with George Kendrick; 835.6363/50, Feb. 25, 1921, Stimson (Buenos Aires) to Secretary of State. Also see Braden, p. 73.

43. *Ibid.*, 835.6363/25, Aug. 16, 1920, Sadler (Standard Oil Co. of New Jersey in New York) to Secretary of State. For Moreno's interview with Demarchi, see 835.6363/38, Oct. 30, 1920, Wadsworth (Buenos Aires) to Secretary of State.

44. United States, [3], 835.6363/79, Sep. 28, 1921. Standard Oil Co. of New Jersey to Secretary of State; Great Britain, [1], 371/7173, Jan. 31, 1922, Macleay (Buenos Aires) to Foreign Office; Stephens, p. 9.

45. Gibb and Knowlton, pp. 381, 646.

46. *The Review of the River Plate*, 52 (Aug. 1, 1919), p. 317; *La Epoca*, Jul. 14, 1920, p. 1.

47. "Yacimientos Petrolíferos Fiscales de Comodoro Rivadavia," *Petróleo y Minas*, 9 (Aug. 1, 1929), p. 7; Argentina, [26], pp. 92–93.

48. Argentina, [18], p. 38.

49. Argentina, [13], p. 4; Argentina, [15], p. 53; "De Comodoro Rivadavia," *El Petróleo Argentino: Revista Quincenal de la Industria Nacional del Petróleo y sus Derivados*, 1 (Aug. 10, 1923), p. 17.

50. Argentina, [14], p. 55; *La Razón*, Dec. 11, 1919 (clipping in Biblioteca Tornquist, Buenos Aires).

51. *La Vanguardia*, Aug. 17, 1919, p. 3; Aug. 25, 1919, p. 3; Aug. 27, 1919, p. 4; Oct. 17, 1919, p. 3; Oct. 23, 1919, p. 5. *The Review of the River Plate*, 52 (Sep. 12, 1919), p. 675; 52 (Oct. 31, 1919), p. 1131. *La Prensa*, Sep. 9, 1919, p. 12. Argentina, [14], p. 5.

52. *La Vanguardia*, Oct. 23, 1919, p. 5.

53. *La Prensa*, Dec. 15, 1919, p. 10; *La Vanguardia*, Feb. 22, 1920, p. 9; *The Review of the River Plate*, 52 (Dec. 21, 1919), p. 1657.

54. *La Prensa*, Jan. 7, 1920, p. 6; Feb. 3, 1920, p. 1; Jan. 17, 1920, p. 6. *La Vanguardia*, Jan. 23, 1920, p. 9; Feb. 27, 1920, p. 9; Jan. 14, 1920, p. 5.

55. *La Protesta*, Jan. 26, 1920, p. 4; Feb. 5, 1920, p. 3. *La Vanguardia*, Jan. 24, 1920, p. 5.

56. Argentina, [15], p. 63. *The Review of the River Plate*, 53 (Mar. 5, 1920), p. 631; 53 (Mar. 26, 1920), p. 835. *La Vanguardia*, Feb. 25, 1920, p. 2; Sep. 12, 1920, p. 2.

57. Argentina, [2], Sesiones ordinarias, vol. 4 (Aug. 24, 1920), p. 452; vol. 4 (Sep. 2, 1920) p. 773.

58. *Ibid.*, vol. 4 (Sep. 7, 1920), pp. 839–49. The quoted material appears on p. 846. Also see *ibid.*, vol. 5 (Sep. 22, 1920), pp. 451, 559.

59. Argentina, [15], pp. 5–7.

60. Argentina, [23], p. 11; Argentina, [4], vol. 1, pp. 172, 227–29.

61. *La Prensa*, Jul. 21, 1922, p. 11; *La Vanguardia*, May 22, 1922, p. 1. For additional press criticism of the government's operation of the state oil fields, see *The Review of the River Plate*, 55 (Jan. 21, 1921), p. 182; *Buenos Aires Herald*, Jul. 28, 1922, enclosure in United States, [3], 835.6363/169, Aug. 3, 1922, Robertson (Buenos Aires) to Secretary of State; and "La explotación fiscal del petróleo en Comodoro Rivadavia: grave situación," *Petróleo y Minas*, 2 (Jun. 15, 1922), p. 16.

62. "Anarquía institucional," *Petróleo y Minas*, 2 (Jan. 15, 1922), p. 13.

63. *The Review of the River Plate*, 56 (Aug. 19, 1921), p. 493; *La Vanguardia*, Jun. 2, 1922, p. 1. For a thorough examination and indictment of the Yrigoyen government's irregular administration of the oil fields, see the document prepared by Deputy Rodolfo Moreno in Argentina, [2], Sesiones ordinarias, vol. 2 (Jun. 1, 1923), pp. 240–53.

64. Guillermo F. Romero, employed at Comodoro Rivadavia in 1921, made these comments to *La Vanguardia*, Sep. 19, 1927, p. 2. Colonel Enrique Mosconi, who assumed command of YPF in 1923, confirmed these observations. See his *El petróleo argentino*, p. 46. Also see *The Review of the River Plate*, 59 (Feb. 9, 1923), p. 352.

65. "En la Dirección General de Minas," *Petróleo y Minas*, 2 (Apr. 15, 1922), p. 5.

66. United States, [3], 835.6363/150, Jun. 13, 1922, Robertson (Buenos Aires) to Secretary of State; 835.002/59, Jun. 1, 1922, Riddle (Buenos Aires) to Secretary of State. *La Vanguardia*, Jun. 2, 1922, p. 1. *The Review of the River Plate*, 57 (Jun. 2, 1922), pp. 1349, 1408.

67. For the text of Yrigoyen's decree creating YPF, see Argentina, [26], pp. 95–97.

68. *Buenos Aires Herald*, Jun. 5, 1922, enclosure in United States, [3], 835.6363/148, Jun. 19, 1922, Robertson (Buenos Aires) to Secretary of State; *La Prensa*, Jun. 5, 1922, p. 5. Also see *La Vanguardia*, Jun. 4, 1922, p. 1, and Jun. 8, 1922, p. 1; and *Quarterly Times of Argentina*, Jun. 5, 1922, enclosure in United States, [3], 835.6363/150, Jun. 13, 1922.

69. Argentina, [2], Sesiones ordinarias, vol. 2 (Jul. 19, 1922) pp. 39–46. The quote is on p. 41.

70. *Ibid.*, vol. 2 (Jul. 19, 1922), pp. 46–72; vol. 2 (Jul. 20, 1922) p. 231; United States, [3], 835.6363/164, Jul. 22, 1922, Robertson (Buenos Aires) to Secretary of State.

71. *La Prensa*, Jul. 21, 1922, p. 11.

72. Argentina, [2], Sesiones ordinarias, vol. 2 (Jul. 26, 1922), p. 368; vol. 2 (Jul. 28, 1922), pp. 436, 451. In the July 28 vote, 54 Radical deputies opposed

the Repetto motion to investigate Yrigoyen's oil policy and 11 Radicals supported it.

73. Argentina, [4].

74. United States, [3], 835.00/266, Jul. 10, 1922, White (Buenos Aires) to Secretary of State. For Salaberry's suicide, see *The Review of the River Plate*, 60 (Nov. 16, 1923), p. 1149.

75. For a review of the case and its antecedents, see the published report of the 1923 judicial proceedings, *Un caso de contrabando y defraudación: Frontini Don Angel Guillermo y Ministerio Fiscal contra "West India Oil Company" y "Compañía Nacional de Petróleos"* (Buenos Aires, 1923), pp. 22–75. Also see Hollander, p. 284.

76. Salaberry's instructions are in another judicial document, *Las grandes defraudaciones: West India Oil Company (WICO), Compañía Nacional de Petróleos: Informaciones sobre el proceso judicial: Demonstración de fraude* (Buenos Aires, 1923), p. 17.

77. United States, [3], 835.00/361, Jul. 30, 1925, Thaw (Buenos Aires) to Secretary of State; "The End of a Six-Year Lawsuit Through All the Courts," *Comments on Argentine Trade*, 7 (May 1928), p. 21.

78. "Hostilidad a las empresas privadas," *Petróleo y Minas*, 8 (Jun. 1, 1928), pp. 10–11; "The End of a Six-Year Lawsuit Through All the Courts," p. 21.

79. *La Prensa*, Jul. 21, 1922, p. 11.

80. United States, [3], 835.6363/147, Jun. 1, 1922, Robertson (Buenos Aires) to Secretary of State.

Notes to Chapter 4

1. Imaz, *Los que mandan*, p. 71.

2. Peter H. Smith, *Politics and Beef in Argentina*, pp. 83–87; Vernon L. Phelps, pp. 43–45, 48–53.

3. *The Review of the River Plate*, 58 (Jul. 28, 1922), p. 207. For the Prince of Wales's visit, see Argentina, [2], Sesiones ordinarias, vol. 3 (Jul. 30, 1925), p. 490, and vol. 3 (Aug. 21, 1925), pp. 602–4. For comment on the economic significance of the visit, see Robertson, p. 228.

4. Luna, *Alvear*, p. 40; Molina, p. 272.

5. Ibarguren, p. 348; Luna, *Alvear*, pp. 42–49.

6. Rock, *Politics in Argentina*, pp. 223–24; Great Britain, [1], 371/7172, Oct. 27, 1922, Mallet (Buenos Aires) to Foreign Office; 371/8146, Jun. 30, 1923, Alston (Buenos Aires) to Foreign Office.

7. Peter H. Smith, *Argentina and the Failure of Democracy*, pp. 78–79; Rock, *Politics in Argentina*, pp. 114, 229–31. For analysis of the dispute over sugar policy, see Solberg, "The Tariff," pp. 267–70.

8. For details on the Radical schism of 1924, see Molina, pp. 290–301; and Rock, *Politics in Argentina*, pp. 229–31.

9. The term "contubernio" refers to an illicit marriage or a despicable alliance. See Coca, p. 8; Bayá, p. 19, Cantón, vol. 1, pp. 43–47.

10. London *Times*, Aug. 16, 1924, p. 1; *The Review of the River Plate*, 64 (Jan. 1, 1926), p. 5; 62 (Oct. 3, 1924), p. 861.

11. For Le Bretón's views on the importance of YPF, see *The Review of the River Plate*, 62 (Dec. 26, 1924) p. 1651. Also see Goodwin, "The Politics of Rate-Making," p. 258.

12. Larra, p. 34. Alvear's decree appointing Mosconi, issued on Oct. 19, 1922, is in Argentina, [26], p. 106.

13. See the model of economic nationalism in Johnson, "The Ideology of Economic Policy in the New States," pp. 129–30.

14. Lugones, pp. 103, 110. Lugones specifically acknowledges Bunge on p. 135. Also see Montenegro, p. 449; Troncoso, pp. 39–42; and Potash, pp. 19, 47.

15. See, for example, Torres, "Aviación: Fomento e industria," p. 71; and Torres, "Aviones metálicos, aviones junkers," p. 588.

16. Baldrich, *El problema del petróleo*, p. 37.

17. Montenegro, pp. 454–64; Torres, "Aviones metálicos, aviones junkers," pp. 583–88.

18. Argentina, [27], pp. 53–54; Argentina, [28], pp. xi–xii.

19. Vicat, "Ideas sueltas," p. 179; Vicat, "Necesidad," p. 130.

20. Guevara, "Favorecer a la industria privada," p. 737; Guevara, "Industria siderúgica," pp. 397–403; Montenegro, pp. 454–64.

21. Barrera, p. 17.

22. Vicat, "Combustibles y defensa nacional" (Sep. 1923), pp. 347, 349.

23. For a summary of Baldrich's arguments on petroleum, see Baldrich, *El petróleo*, pp. 14–20.

24. Barcía Trelles, pp. 7–39, 210–28, 248.

25. Baldrich, "El petróleo. Su importancia comercial, industrial y militar. Legislación petrolera," in *El petróleo. Su importancia en la vida de las naciones . . . ,* p. 53.

26. My analysis of the navy's position on oil development is somewhat speculative, since the library of the *Club Naval* in Buenos Aires, whose holdings might have shed light on this matter, was barred to me.

27. For comparisons of Mosconi with the giants of the international oil industry, see Larra, p. 80, and Cáceres Cano, p. 28. On Enrico Mattei and ENI, see Frankel, pp. 61–111; and Votaw, pp. 1–22.

28. Larra, pp. 14–18; Guevara Labal, *El General Ingeniero Enrique Mosconi*, pp. 21–22. For Mosconi's remarks on his service in Germany, see Mosconi, *Dichos y hechos*, p. 180.

29. Guevara Labal, *El General Ingeniero Enrique Mosconi*, pp. 21–22.

30. Mosconi, *Dichos y hechos*, p. 34. Also see Larra, pp. 22–23, 195.

31. Mosconi, *Dichos y hechos*, pp. 94–95; Great Britain, [1], 371/5521, Mar. 7, 1921, Macleay (Buenos Aires) to Foreign Office, p. 26.

32. Mosconi, *El petróleo argentino*, p. 15; Mosconi, *Creación*, p. 155.

33. United States, [3], 835.6363/187, Nov. 21, 1922, Riddle (Buenos Aires) to Secretary of State; Larra, p. 115.

34. Buenos Aires *Standard*, Feb. 7, 1923, enclosure in United States, [3],

835.6363/204; United States, [3], 835.6363/187, Nov. 21, 1922, Riddle (Buenos Aires) to Secretary of State; Great Britain, [1], 371/5521, Mar. 7, 1921, Macleay (Buenos Aires) to Foreign Office; Larra, pp. 34, 44.

35. Mosconi, *El petróleo argentino*, p. 63; Larra, p. 56.

36. "Coronel Enrique Mosconi," *Petróleo y Minas*, 2 (Nov. 15, 1922), p. 6. Mosconi reviews YPF's situation when he assumed control in his *El petróleo argentino*, pp. 35–50.

37. Mosconi, *El petróleo argentino*, p. 50.

38. Alvear's decree, which implemented the new YPF organizational structure, is found in Argentina, [25], pp. 3–17. For a summary, see *The Review of the River Plate*, 59 (Apr. 20, 1923), p. 973.

39. Mosconi, *Dichos y hechos*, p. 151.

40. Argentina, [24], pp. 41, 48, foldouts facing pp. 64, 70. The plan is summarized in Rumbo, pp. 50–51. Also see Mosconi, *El petróleo argentino*, pp. 57–62, 82.

41. Mosconi, *El petróleo argentino*, p. 143–45; Guevara Labal, *El petróleo y sus derivados*, p. 158; Argentina, [24], p. 32.

42. Larra, p. 78; Mosconi, *El petróleo argentino*, p. 117.

43. Mosconi, *El petróleo argentino*, pp. 116–17; Argentina, [2], Sesiones ordinarias, vol. 3 (Jun. 12, 1923), pp. 297–98; vol. 4 (Jun. 27, 1923), pp. 510–20. For Alvear's preliminary contract with Bethlehem, see Argentina, [26], pp. 140–42; for the financing arrangements, see pp. 138–40, 192–93.

44. Great Britain, [1], 371/10605, Apr. 8, 1925, Alston (Buenos Aires) to Foreign Office; May 5, 1925, Alston (Buenos Aires) to Foreign Office. Wilkins, "Multinational Oil Companies," p. 424.

45. "Destilería fiscal del petróleo," *Petróleo y Minas*, 6 (Jan. 1, 1926), p. 9; Great Britain, [1], 371/10605, Apr. 8, 1925, Alston (Buenos Aires) to Foreign Office; Buchanan, p. 159.

46. "La destilería de La Plata," *Petróleo y Minas*, 7 (Sep. 1, 1927), p. 14; Mosconi, *El petróleo argentino*, p. 119; Buchanan, pp. 159–60.

47. Larra, p. 44; Craviotto, p. 555; Argentina, [34], p. 365.

48. Argentina, [34], pp. 365–74.

49. For the contract with Auger, see Argentina, [19], pp. 8–9; Argentina, [12], p. 21. Also see Cochran and Reina, pp. 54–55.

50. "Resumen estadístico de la economía argentina," *Revista de Economía Argentina*, 20 (Nov. 1936), p. 325.

51. Serghiesco, pp. 8–13. For a further comment on YPF's production problems, see "El petróleo nacional," *Petróleo y Minas*, 9 (Apr. 1, 1929), p. 8; and "La conferencia del Sr. Colombo," *Petróleo y Minas*, 7 (Jun. 1, 1927), p. 19.

52. Argentina, [21], pp. 16–20; Mosconi, *El petróleo argentino*, p. 7.

53. Bunge, *The Cost of Living*, p. 8. *La Prensa*, Dec. 6, 1929, p. 15; "Lo de siempre," *Petróleo y Minas*, 10 (Apr. 1, 1930), p. 13; *Financial Times* (London), Mar. 6, 1923, enclosure in United States, [3], 835.6363/205, Mar. 7, 1923, De Vault (London) to Secretary of State.

54. Argentina, [34], p. 93; Cáceres Cano, pp. 65, 73. "Lo de siempre," *Pe-*

tróleo y Minas, 10 (Apr. 1, 1930), p. 13; Mosconi, *El petróleo argentino*, pp. 66–69, 72–73.

55. Guevara Labal, *El General Ingeniero Enrique Mosconi*, p. 81; Argentina, [20], pp. 9, 42.

56. Mosconi, *El petróleo argentino*, pp. 187–88.

57. "El nuevo administrador de los Yacimientos Fiscales," *El Petróleo Argentino: Revista Quincenal de la Industria Nacional del Petróleo y sus Derivados*, 1 (May 25, 1923), p. 10; "De Comodoro Rivadavia," in *ibid.*, (Aug. 25, 1923), pp. 17–18; "La administración local de Y.P.F. en C. Rivadavia," in *ibid.*, (Sep. 25, 1923), pp. 9–10. See also *La Vanguardia*, Sep. 19, 1927, p. 3.

58. *The Review of the River Plate*, 62 (Aug. 1, 1924), p. 279. "Zonas en explotación de Comodoro Rivadavia," *Petróleo y Minas*, 4 (Aug. 15, 1924), p. 13. *La Protesta*, Jul. 30, 1924, p. 2; Jul. 31, 1924, p. 2; Aug. 10, 1924, p. 1.

59. *La Protesta*, Sep. 30, 1927, p. 2.

60. For chronicles of this strike, see *La Protesta*, Sep. 24, 1927, p. 2; Aug. 16, 1927, p. 1. *La Vanguardia*, Sep. 21, 1928, p. 3.

61. *La Protesta*, Aug. 16, 1927, p. 1; Aug. 19, 1927, p. 1.

62. Alvear's proposed legislation is in Argentina, [2], Sesiones ordinarias, vol. 4 (Sep. 20, 1923), pp. 244–45.

63. *Ibid.*, vol. 4 (Sep. 10, 1925), pp. 306–7; *The Review of the River Plate*, 63 (Sep. 18, 1925), p. 41; Coca, p. 39.

64. For the decrees, see Argentina, [26], pp. 145–51. For analysis of their content and of Alvear's decision to issue them, see *The Review of the River Plate*, 61 (Jan. 18, 1924), p. 171, and 61 (May 2, 1924), p. 1095; Dudley M. Phelps, "Petroleum Regulation," p. 51; and Hileman, *Sobre legislación*, p. 23.

65. Argentina, [2], Sesiones ordinarias, vol. 3 (Aug. 10, 1926), p. 575; and [Unión Industrial Argentina], *Estado actual*, p. 6.

66. "Las perforaciones en los cateos," *Petróleo y Minas*, 4 (May 15, 1924), p. 1; "La minería y el mensaje presidencial," in *ibid.*, (Jul. 15, 1924), p. 11; "La evolución industrial," in *ibid.*, 5 (Jul. 15, 1925), p. 6; "Apuntes parlamentarias," in *ibid.*, 4 (Aug. 15, 1924), p. 8.

67. [Unión Industrial Argentina], *Estado actual*, p. 6.

68. Hileman, *Sobre legislación*, p. 15; Lagos, *El petróleo*, pp. 21–24.

69. Great Britain, [1], 371/10605, Apr. 8, 1925, Alston (Buenos Aires) to Foreign Office; London *Times*, Apr. 4, 1925, p. 11.

70. Correspondence between Sir Charles Greenway and the Minister of Agriculture is found in Great Britain, [1], 371/10605, Apr. 8, 1925, Alston (Buenos Aires) to Foreign Office.

71. *The Review of the River Plate*, 63 (Mar. 27, 1925), p. 7; Great Britain, [1], 371/10605, May 5, 1925, Alston (Buenos Aires) to Foreign Office; London *Times*, Apr. 4, 1925, p. 11; *Buenos Aires Herald*, Mar. 21, 1925, p. 1.

72. United States, [3], 835.6363/247, Sep. 11, 1923, Robertson (Buenos Aires) to Secretary of State; 835.6363/222, May 27, 1923, Robertson (Buenos Aires) to Secretary of State.

73. Bucich Escobar, pp. 500–501; United States, [3], 835.00/365, Sep. 10,

1925, Thaw (Buenos Aires) to Secretary of State; 835.00/370, Nov. 5, 1925, Thaw (Buenos Aires) to Secretary of State. Great Britain, [1], 371/11112, Feb. 8, 1926, Robertson (Buenos Aires) to Foreign Office. Molina, pp. 302–9.

74. For an overview of Salta's economy and of the history of the province's oil policy, see Hollander, pp. 151–72 and 312–39. Also see Caro Figueroa, pp. 113–22. It is interesting that Yrigoyen permitted liberal mineral-land acquisition procedures in Salta in 1918 only a year before he introduced legislation in Congress to control provincial petroleum concession policy. I am not able to explain this seeming anomaly entirely, but I believe that during the first two years of his presidency, about through the end of 1918, Yrigoyen subordinated national petroleum policy to the interests of maintaining the fragile regional unity of the Radical Party (see the evidence in Chapter 2). This attempt to conciliate dissident Radical opinion in the interior provinces may explain his liberal attitude toward Salta's oil concessions during the 1918 intervention. Then, as urban opinion mounted against the power of the oil "trusts," especially during and after the 1919 congressional antitrust investigations, Yrigoyen introduced legislation to control provincial petroleum policy. By this time, the Radical Party in the northwestern provinces was beginning to reject the President's leadership, largely because of his sugar policy, a crucial issue for the region. In other words, as the party began to split because of economic conflict and along regional lines, Yrigoyen abandoned his earlier conciliatory petroleum policy in the provinces.

75. Jaime, p. 160; Bonardi, pp. 2547–48; Hollander, pp. 183–85, 353–72.

76. The Review of the River Plate, 59 (Jul. 27, 1923), p. 201; "Pozos de petróleo en Jujuy," Petróleo y Minas, 6 (Aug. 1, 1926), p. 22. For Villafañe's decrees, see Argentina, [26], pp. 165–69.

77. For a review of Mosconi's 1924 campaign against Standard, see Argentina, [19], p. 6; "En las provincias," Petróleo y Minas, 5 (Jun. 15, 1925), p. 19; and Mosconi, El petróleo argentino, pp. 191–92. For a review of these events by an agent of Standard Oil, see the report of Carl Kincaid in United States, [3], 835.6363/292, Oct. 18, 1926. For analysis of the background to Standard's Salta concessions, see Augusto Bunge, pp. 73–74, 81–83.

78. United States, [3], 835.6363/292, Nov. 11, 1926, Sadler (Standard Oil of N.J.) to Secretary of State. For a detailed account of Standard's operations in Salta, see La Nación, Dec. 13, 1928, Third Section, p. 17. For Corbalán's position on the 1924 decrees, see Argentina, [2], Sesiones ordinarias, vol. 6 (Dec. 2, 1926), p. 734. Also see Hollander, pp. 382–84, 427.

79. La Nación, Dec. 13, 1928, Third Section, p. 17; "Petroleum in Salta," The Review of the River Plate, 66 (Apr. 20, 1928), p. 19.

80. Mosconi's note is in Argentina, [2], Sesiones ordinarias, vol. 5 (Sep. 9, 1926), p. 299.

81. Correspondence among Mosconi, the ministers of Agriculture and the Interior, and the provincial governors is in "Documentación de un incidente," Petróleo y Minas, 6 (Oct. 1, 1926), pp. 4–19. Also see Argentina, [2], Sesiones ordinarias, vol. 6 (Sep. 22, 1926), p. 213.

82. Villafañe, ed., El petróleo, pp. 7–8; and Villafañe, "Gravedad," pp. 36–38.

83. Mosconi's message is in Argentina, [2], Sesiones ordinarias, vol. 3 (Aug. 10, 1926), pp. 610–13. Also see "Pozos de petróleo en Jujuy," *Petróleo y Minas*, 6 (Aug. 1, 1926), p. 22; and *The Review of the River Plate*, 64 (Nov. 5, 1926), p. 41, and 64 (Dec. 3, 1926), p. 32. For Alvear's decree, see Argentina, [26], pp. 227–28.

84. For this interchange of correspondence, see Argentina, [2], Sesiones ordinarias, vol. 6 (Dec. 2, 1926), pp. 704–6. Also see *El Intransigente*, Oct. 16, 1926, quoted in *El petróleo del norte argentino*, p. 112.

85. United States, [3], 835.6363/292, Nov. 11, 1926, Sadler (New York) to Secretary of State.

86. *La Epoca*, Oct. 23, 1926, p. 1; and Dec. 30, 1926, p. 1. Also see *The Review of the River Plate*, 64 (Aug. 6, 1926), p. 7.

87. Villafañe, *La región de las parias*, pp. 56–57. His two attacks on Yrigoyen were *El yrigoyenismo: No es un partido político. Es una enfermedad nacional y un peligro público* and *Degenerados: Tiempos en que la mentira y el robo engendrán apóstoles*.

88. Argentina, [20], p. 6; Argentina, [26], p. 263.

89. This affair is related in Lencinas, pp. 7, 14. For background on *lencinismo*, see Rodríguez, pp. 383–89, 411–27.

90. Craviotto, p. 551; Argentina, [34], p. 155.

Notes to Chapter 5

1. The motor-vehicle trade illustrates the growing American predominance in the Argentine market. In 1927, the U.S. exported 58,745 cars and trucks to Argentina, whereas Italy, in second place, sent 927, France sent 796, and Britain sent 758! See Hipwell, p. 153. For further background on the rapid growth of U.S. exports to Argentina, see Fodor and O'Connell, p. 18, and Dye, p. 21.

2. Gravil, p. 50; Wilkins, *The Maturing of Multinational Enterprise*, p. 61; London *Times*, Mar. 4, 1930, p. 15; Great Britain, [1], 371/14196, May 24, 1930, Millington-Drake (Buenos Aires) to Foreign Office.

3. Sweet, pp. 46–54; Gravil, p. 45.

4. *The Review of the River Plate*, 61 (Mar. 28, 1924), p. 747; United States, [3], 835.00/368, Oct. 8, 1925, Thaw (Buenos Aires) to Secretary of State.

5. Duhau, "Fomento," p. 10.

6. Machado, p. 5; Sweet, p. 108.

7. *La Prensa*, Mar. 10, 1927, p. 12. Also see United States, [3], 835.00/390, Jan. 11, 1927, Cable (Buenos Aires) to Secretary of State.

8. Lewis, pp. 41, 83–84; Russell, p. 111; Benham, pp. 76–78.

9. Great Britain, [1], 395/420, Jul. 31, 1927, Robertson (Buenos Aires) to Foreign Office. Also see Fodor and O'Connell, p. 36; and United States, [3], 835.00/392, Mar. 9, 1927, Cable (Buenos Aires) to Secretary of State.

10. Duhau, "Intercambio," p. 1006; "Comprar a quien nos compra," *Anales de la Sociedad Rural Argentina*, 60 (Mar. 15, 1927), pp. 265–67; "Estados Unidos prohibe la importación de carnes argentinos," *ibid.* (Jan. 1, 1927), pp. 17–18.

11. For analysis of the anti-American content of the Argentine press in 1927,

see Kurtz, pp. 227–39. Also see Easum, p. 109; and *La Prensa*, Feb. 22, 1927, p. 9.

12. *La Vanguardia*, Aug. 7, 1927, p. 1; Aug. 11–12, 1927, p. 1. Marotta, vol. 3, pp. 243–45.

13. United States, [3], 835.00/B8, Dec. 26, 1927, Bliss (Buenos Aires) to Secretary of State; 835.6363/309, Feb. 15, 1928, Bliss (Buenos Aires) to Secretary of State. Alarmed at the rise of anti-Americanism, observers in the U.S. embassy reported that "certain British elements in Buenos Aires" were subsidizing the most hostile newspapers. The Foreign Office correspondence lends some support to this observation. See United States, [3], 835.00/391, Feb. 10, 1927, Cable (Buenos Aires) to Secretary of State; and Great Britain, [1], 395/420, Jul. 31, 1927, Robertson (Buenos Aires) to Foreign Office.

14. Rock, *Politics in Argentina*, pp. 232–39; Frondizi, *Petróleo y política*, pp. 194–95.

15. Quoted in *El petróleo del norte argentino*, p. 71.

16. Rock, *Politics in Argentina*, pp. 21–24; Rock, "Radical Populism," pp. 69–70.

17. Johnson, "A Theoretical Model," pp. 4–5, 14–15.

18. Pearton, p. 189. For representative petitions from student organizations in support of the petroleum monopoly, see Argentina, [2], Sesiones ordinarias, vol. 1 (Jun. 22, 1927), p. 710; vol. 2 (Jul. 14, 1927), p. 218; vol. 3 (Aug. 10, 1927), p. 513. Also see "Los cursos sobre explotación e industrialización del petróleo," *Boletín de Informaciones Petrolíferas*, 3 (Jul. 1926), p. 819; and Walter, *Student Politics in Argentina*, pp. 9–10, 71–73.

19. For an analysis of the theoretical relationships between populist movements and the social class structure in countries at an intermediate level of development, see Van Niekerk, pp. 189–92.

20. *La Epoca*, Jul. 27, 1927, p. 1; Jul. 28, 1927, p. 1; and Jul. 31, 1927, p. 1.

21. *La Epoca*, Aug. 3, 1927, p. 1. Also see issues of Aug. 8, 1927, p. 1; Aug. 12, 1927, p. 1; and Aug. 17, 1927, p. 1.

22. United States, [3], 835.6363/296, Jun. 28, 1927, Cable (Buenos Aires) to Secretary of State. For Baldrich's speech at the Centro Naval on Feb. 2, 1927, see Baldrich, *El petróleo*, p. 30.

23. *La Prensa*, Jul. 24, 1927, p. 19; Jul. 26, 1927, p. 13.

24. Baldrich, *El problema del petróleo*, pp. 40–41.

25. Standard's statement appeared in "Manifestación del honor," *Petróleo y Minas*, 7 (Sep. 1, 1927), pp. 11–13.

26. For the private companies' statement, see [Unión Industrial Argentina], *Estado actual*. Also see Argentina, [2], Sesiones ordinarias, vol. 2 (Jul. 14, 1927), pp. 220–37; and "La política del petróleo," *Petróleo y Minas*, 7 (Jul. 1, 1927), pp. 14–15.

27. Cornejo, pp. 99–105; Villafañe, ed., *El petróleo*, pp. 19, 25; Lencinas, p. 9. Also see Correa, pp. 262–63.

28. Sánchez Sorondo, pp. 65–66.

29. *Ibid.*, pp. 67, 71, 77. Sánchez Sorondo's reference to rents apparently al-

luded to a temporary rent freeze in the Federal Capital that Yrigoyen induced Congress to approve in 1920.

30. Colombo, *El petróleo argentino*, pp. 4, 15. For the UIA's petition to Congress in support of its position, see Argentina, [2], Sesiones ordinarias, vol. 2 (Jul. 20, 1927), p. 398. Also see *La Prensa*, Jul. 22, 1927, p. 9; Jul. 23, 1927, p. 9.

31. For an analysis of the Socialist schism and the decision of the PSI to support the petroleum monopoly, see Coca, pp. 62–63; and Walter, *The Socialist Party of Argentina*, pp. 205–10.

32. *La Vanguardia*, Aug. 21, 1927, p. 1. Also see Sep. 10, 1927, p. 1; and Sep. 13, 1927, p. 1.

33. Argentina, [2], Sesiones ordinarias, vol. 2 (Jun. 24, 1927), p. 69.

34. *Ibid.* (Jul. 20, 1927), p. 498; *The Review of the River Plate*, 65 (Aug. 12, 1927), p. 13. Also see Frondizi, *Petróleo y política*, pp. 195–96.

35. For reports on Molinari, see United States, [3], 835.00/266, Jul. 10, 1922, White (Buenos Aires) to Secretary of State; and Great Britain [1], 371/14195, Jan. 30, 1930, Millington-Drake (Buenos Aires) to Foreign Office.

36. Argentina, [2], Sesiones ordinarias, vol. 3 (Jul. 28, 1927), pp. 184–209.

37. *Ibid.* (Jul. 28, 1927), pp. 219–96; (Jul. 29, 1927), pp. 326–33; and vol. 4 (Sep. 1, 1927), p. 327.

38. *Ibid.*, vol. 3 (Aug. 11, 1927), pp. 670–87.

39. *Ibid.* (Aug. 4, 1927), p. 459.

40. *Ibid.* (Aug. 10, 1927), pp. 617–31.

41. *Ibid.* (Aug. 3, 1927), p. 421; (Aug. 17, 1927), pp. 795–96; (Aug. 18, 1927), pp. 830–39. Also see Lencinas, p. 6.

42. Argentina, [2], Sesiones ordinarias, vol. 4 (Sep. 1, 1927), pp. 334, 362–63.

43. "Los efectos," *Petróleo y Minas*, 7 (Nov. 1, 1927), p. 7.

44. London *Times*, Jul. 5, 1929, p. 15; *La Epoca*, Feb. 21, 1928, p. 1.

45. Rock, *Politics in Argentina*, p. 240.

46. For the election results, see Etchepareborda, "La segundo presidencia de Hipólito Yrigoyen," pp. 354–55.

47. *Buenos Aires Herald*, Apr. 13, 1928, enclosure in United States, [3], 835.00/423, Bliss (Buenos Aires) to Secretary of State.

48. *La Prensa*, Sep. 9, 1928, p. 12; *La Protesta*, Sep. 7, 1928, p. 1; Alejandro E. Bunge, "La legislación del petróleo," *Revista de Economía Argentina*, 21 (Oct. 1928), p. 289; Lencinas, p. 9.

49. *La Epoca*, Aug. 31, 1928, p. 1; Sep. 16, 1928, p. 1.

50. For the debates, see Argentina, [2], Sesiones ordinarias, vol. 4 (Sep. 17, 1928), pp. 357–89.

51. Argentina, [5], Sesiones ordinarias, vol. 1 (Sep. 24, 1928), pp. 691–95; *La Epoca*, Aug. 31, 1928, p. 1.

52. Argentina, [2], Sesiones ordinarias, vol. 4 (Sep. 7, 1928), p. 362.

53. United States, [3], 835.6363/309, Feb. 15, 1928, Bliss (Buenos Aires) to Secretary of State; Mosconi *El petroleo argentino*, p. 216.

54. The complete text of the Mosconi address at the National University is in United States, [3], 835.6363/311, May 15, 1928, Armstrong (Standard Oil of New Jersey) to Morgan (Department of State). Also see *Excelsior* (Mexico City), Feb. 9, 1928, p. 4; Larra, pp. 61–66; and Etchepareborda, "Notas sobre la administración de Marcelo T. de Alvear," pp. 144–45.

55. Mosconi, *El petróleo argentino*, pp. 227–30; Randall, pp. 184–87.

56. Mosconi, *El petróleo argentino*, pp. 230–31; Puga Vega, pp. 73–75.

57. Mosconi, *Dichos y hechos*, pp. 127–29, 136–38, 198, 209–10.

58. *Ibid.*, p. 137; Mosconi, *La batalla del petróleo*, pp. 116–17. Also see Mosconi, "Prólogo," p. viii.

59. Quoted in Gibb and Knowlton, p. 382.

60. Mosconi, *El petróleo argentino*, pp. 181–82; *idem*, "Prólogo," p. xxx; and *idem*, *Dichos y hechos*, p. 217.

61. This warning appears in a letter from Mosconi to Ricardo Oneto of June 27, 1929, and is cited in Oneto, p. xv.

62. Oneto, p. xvi. Also see United States, [3], 835.6363/333, Oct. 7, 1929, Bliss (Buenos Aires) to Secretary of State; "El monopolio de petróleo," *Petróleo y Minas*, 8 (Apr. 1, 1928), p. 17. For Mosconi's interest in the Anglo-Persian organizational structure, see Palacios, p. 100.

63. Guevara Labal, *El petróleo*, p. 90; Dorfman, p. 148.

64. Argentina, [12], pp. 17–25.

65. Serghiesco, pp. 10–13.

66. *Ibid.*, p. 15; United States, [3], 835.6363/335, Nov. 14, 1929, Bliss (Buenos Aires) to Secretary of State; "Roedor de millones," *Petróleo y Minas*, 9 (Sep. 1, 1929), pp. 14–15.

67. Great Britain, [1], 371/14196, May 24, 1930, Millington-Drake (Buenos Aires) to Foreign Office; "Foreign Producers in Argentina Facing Uncertain Government Policy," *Oil Weekly*, 55 (Dec. 13, 1929), p. 73.

68. For comparison of Argentine gasoline prices with those in other Latin American countries in 1930, see Mosconi, *El petróleo argentino*, p. 159. Also see Casella and Clara, p. 17.

69. Larson, Knowlton, and Popple, pp. 305–8; and Sampson, pp. 70–71.

70. Tugendhat, pp. 93–102; Hartshorn, p. 157; Penrose, pp. 179–82; Sampson, pp. 72–73.

71. For an analysis of Soviet petroleum exports during this period, see Hassman, pp. 54–55. Also see United States, [3], 835.00/375, Jan. 14, 1926, Jay (Buenos Aires) to Secretary of State; 835.00 General Conditions/10, Aug. 22, 1928, Bliss (Buenos Aires) to Secretary of State; Larra, p. 124.

72. Sociedad Anónima "Iuyamtorg," *Exposición-Feria, 1928, Buenos Aires*, p. 13; Great Britain, [1], 371/14196, May 24, 1930, Millington-Drake (Buenos Aires) to Foreign Office.

73. Goldwert, pp. 115–16; Frondizi, *Petróleo y política*, p. 251; Dudley M. Phelps, *The Migration of Industry to South America*, p. 39. In 1929, Argentine import statistics recorded gasoline imports worth 359,777 gold pesos and kerosene

imports worth 22,111 gold pesos from the Soviet Union. Imports for 1930 totaled 55,522 gold pesos of gasoline and 17,121 of kerosene. Argentina, [8], pp. 508–11.

74. At the important coastal city of Mar del Plata, for example, the price of gasoline fell from 28 to 20 centavos, or 29 percent. For detailed analysis of the price reductions, see Argentina, [22], pp. 32–33.

75. Mosconi, *El petróleo argentino*, pp. 169, 173; *The Review of the River Plate*, 67 (Nov. 29, 1929), p. 5.

76. *The Review of the River Plate*, 68 (May 2, 1930), pp. 9, 23. For a similar report by the Compañía Argentina de Comodoro Rivadavia, see *Ibid*. (Mar. 7, 1930), p. 38.

77. Argentina, [21], p. 105.

78. Dickmann, Iñigo Carrera, and Rubenstein, pp. 59, 86–90.

79. Argentina, [22], p. 34. For Cantilo's 1928 decree, see "El Intendente Municipal Dr. Cantilo ha adoptado una medida favorable a la venta de nafta fiscal," *Boletín de Informaciones Petrolíferos*, 7 (Mar. 1930), pp. 269–70. For the 1930 decree, see *La Epoca*, Feb. 17, 1930, p. 5.

80. Guevara Labal, *El petróleo*, p. 110. For WICO's expansion, see "La West India Oil Company en la Argentina," *Petróleo y Minas*, 9 (Jul. 1, 1929), pp. 16–17; and "West India Service Station is New Landmark on Tigre Highway," *Comments on Argentine Trade*, 9 (Dec. 1929), pp. 36–37. *La Argentina*, May 5, 1930, pp. 3, 7.

81. Hollander, pp. 422–26, 436. For Cornejo's decrees, see Argentina, [26], pp. 292–94, 297. Also see United States, [3], 835.00/13, Dec. 11, 1928, Bliss (Buenos Aires) to Secretary of State.

82. *The Review of the River Plate*, 67 (Dec. 20, 1929), p. 27; "También en Salta," *Petróleo y Minas*, 10 (Jun. 1, 1930), p. 4.

83. *La Epoca*, Oct. 25, 1929, p. 1, reported Bonardi's opening address. Bonardi also published his presentation in a series of articles. See "La cuestión petrolífera en Salta," *Revista de Ciencias Económicas*, 31 (Nov. 1928), pp. 2539–50; 31 (Dec. 1928), pp. 2630–44; and 32 (Jan. 1929), pp. 51–68.

84. *La Epoca*, Oct. 31, 1929, p. 2.

85. Naón published his argument in *Inviolabilidad de la propiedad minera*. See pp. 35–36, 48. Also see *La Epoca*, Oct. 31, 1929, p. 2.

86. "Nuestro triunfo," *Petróleo y Minas*, 10 (Jul. 1, 1930), p. 2.

87. *La Epoca*, Oct. 19, 1929, p. 1; Nov. 10, 1929, p. 1. *La Argentina*, Nov. 12, 1929, p. 1. Comité Universitario Radical, Junta Central, *El petróleo argentino*, pp. 1–2.

88. *La Argentina*, Dec. 19, 1929, p. 1.

89. "La Alianza Continental," *Petróleo y Minas*, 9 (Feb. 1, 1929), p. 14; Buchanan, p. 188; Orzábal Quintana, pp. 162–87.

90. United States, [3], 835.6363/303, Oct. 21, 1927, Anderson (Navy Dept.) to Secretary of State.

91. Mosconi, *El petróleo argentino*, pp. 235–36.

92. United States, [3], 835.6363/331, Jun. 12, 1929, White (Buenos Aires) to Secretary of State; *La Epoca*, Nov. 25, 1929, p. 1; *La Vanguardia*, Nov. 28, 1929, p. 1. U.S. observers viewed the Alianza's efforts with some alarm. See United States, [3], 835.6363/336, Dec. 12, 1929, Bliss (Buenos Aires) to Secretary of State; and Nelson, p. 90.

93. For detailed background on *lencinismo* and *cantonismo*, see Rodríguez, pp. 45–104, 184–250, 318–96, and 411–40. Also see Robert Richard Smith, p. 235; Rock, *Politics in Argentina*, pp. 248–50; and Great Britain, [1], 371/13464, Apr. 16, 1929, Robertson (Buenos Aires) to Foreign Office.

94. Argentina, [5], Sesiones ordinarias, vol. 1 (Sep. 27, 1929), p. 649; Sesiones extraordinarias, vol. 3 (Dec. 12, 1929), p. 139.

95. *La Prensa*, Dec. 6, 1929, p. 15; Alejandro E. Bunge, "El problema económico del petróleo," pp. 402, 433.

96. *La Argentina*, Dec. 31, 1929, p. 1; *La Epoca*, Dec. 20, 1929, p. 1.

97. Goodwin, "The Politics of Rate-Making," p. 287.

98. Robertson, p. 226; London *Times*, Apr. 7, 1930, p. 15. Also see United States, [3], 835.51, Dec. 26, 1929, Bliss (Buenos Aires) to Secretary of State.

99. London *Times*, Aug. 2, 1929, p. 13; Great Britain, [1], 371/14196, May 24, 1930, Millington-Drake (Buenos Aires) to Foreign Office, p. 3.

100. Gravil, pp. 54–58; London *Times*, May 24, 1929, p. 13; Oct. 4, 1929, p. 13; Great Britain, [1], 371/14196, May 24, 1930, Millington-Drake (Buenos Aires) to Foreign Office.

101. Quoted in the London *Times*, Jul. 24, 1930, p. 14.

102. For an analysis of the negotiations, see Great Britain, [1], 371/14196, May 24, 1930, Millington-Drake (Buenos Aires) to Foreign Office. The complete text of the trade agreement is in Argentina, [2], Sesiones extraordinarias, vol. 4 (Dec. 4, 1929), pp. 250–51. On the "Silk Rebate Decree," see *The Review of the River Plate*, 67 (Nov. 29, 1929), p. 7. Between 1912 and 1929, the Argentine State Railways purchased 223 locomotives, of which 173 came from the United States. British bids were consistently higher than American ones. United States, [3], 835.77/165, Feb. 21, 1929, S. Walter Washington (Buenos Aires) to Secretary of State. British "artificial silk" was also not competitive in export markets. See the London *Times*, Dec. 15, 1927, p. 17.

103. Robertson, p. 228. The italics are in the original.

104. *La Prensa*, Dec. 25, 1929, p. 5; *The Review of the River Plate*, 68 (Feb. 14, 1930), p. 21; Argentina, [2], Sesiones extraordinarias, vol. 4 (Dec. 12, 1929), p. 474. The D'Abernon pact aroused hostility in Canada, where the press complained of British preference for Argentine over Canadian wheat. See the *Winnipeg Free Press*, Apr. 2, 1930, p. 15; and the *Financial Post* (Toronto), Apr. 3, 1930, p. 1. British farmers also bitterly opposed the agreement. See Great Britain, [2], 233 *H.C. Deb. 5s*, Oct. 30, 1929, p. 165.

105. Great Britain, [1], 371/13464, Sep. 17, 1929, Lindsay (Buenos Aires) to Foreign Office.

106. *Ibid.*, Sep. 20, 1929, Foreign Office to Robertson (Buenos Aires);

371/14191, Jan. 13, 1930, Millington-Drake (Buenos Aires) to Foreign Office; *La Epoca*, Jan. 10, 1930, p. 1; "La nueva destilería, *Petróleo y Minas*, 10 (Jun. 1, 1930), p. 17.

107. *La Epoca*, Nov. 13, 1929, p. 1; Mar. 2, 1930, p. 4. *La Argentina*, Nov. 13, 1929, p. 1.

108. *La Argentina*, Feb. 21, 1930, p. 1; Feb. 22, 1930, p. 1; Feb. 24, 1930, p. 1. Nine of this newspaper's 14 front-page headlines during the two weeks prior to the elections called attention to the oil issue, either to attack Standard or to praise YPF.

109. *La Argentina*, Jan. 28, 1930, p. 3; Feb. 18, 1930, p. 2; Feb. 25, 1930, p. 1; Feb. 26, 1930, p. 1. *La Epoca*, Feb. 2, 1930, p. 2; Feb. 7, 1930, p. 2. Buchanan, p. 284.

110. *Buenos Aires Herald*, Sep. 17, 1922, enclosure in United States, [3], 835.6363/176, Sep. 21, 1922, Robertson (Buenos Aires) to Secretary of State; 835.6363/284, Mar. 7, 1925, Becú (Buenos Aires) to Eskesen (Standard Oil of Argentina, Buenos Aires); Rout, pp. 46–47, 55–57.

111. United States, [3], 835.00 General Conditions/17, Apr. 18, 1929, Bliss (Buenos Aires) to Secretary of State; 835.6363/335, Nov. 14, 1929, Bliss (Buenos Aires) to Secretary of State; Mosconi, *El petróleo argentino*, pp. 194–95.

112. *La Argentina*, Mar. 9, 1930, p. 3; Mar. 10, 1930, p. 1. In 1934, convinced that Standard of Bolivia's desire for a pipeline to the Paraguay River was a major cause of the Chaco War, Baldrich wrote that the 1929 proposal aimed to "convert a belt of Argentine territory into a Polish corridor." Baldrich, *El problema del petróleo*, p. 30.

113. *La Argentina*, Mar. 30, 1930, p. 1. The problem of an exit for Bolivian oil long continued to perturb Argentine-Bolivian relations.

114. For analysis of the failure of petroleum nationalism as a 1930 campaign issue, see Mayo, Andino, and García Molina, p. 163.

115. Vernon L. Phelps, pp. 91–94; Salera, pp. 40–41; Rock, *Politics in Argentina*, pp. 242–43, 255–61; Great Britain, [1], 371/14191, Jun. 28, 1930, Macleay (Buenos Aires) to Foreign Office.

116. Gálvez, pp. 404–5. Great Britain, [1], 371/14196, May 24, 1930, Millington-Drake (Buenos Aires) to Foreign Office; 371/13464, Robertson (Buenos Aires) to Foreign Office.

117. Gálvez, p. 412.

118. Great Britain, [1], 371/14196, May 24, 1930, Millington-Drake (Buenos Aires) to Foreign Office.

119. "Grain Elevators," *The Review of the River Plate*, 65 (Feb. 25, 1927), p. 11; Duhau, *Los elevadores*, p. 46; Confederación Argentina del Comercio . . . , *Actas de la Tercera Conferencia Económica Nacional*, pp. 96–97.

120. *The Review of the River Plate*, 69 (Aug. 29, 1930), p. 13.

121. For an overview of Yrigoyen's agrarian policies during his second government, see Solberg, "Rural Unrest and Agrarian Policy in Argentina, 1912–1930," pp. 50–52.

122. *The Review of the River Plate*, 67 (Feb. 1, 1929), p. 5; Colombo, *Levántate y anda*. One observer evaluated the UIA's campaign as "the strongest organized attack on Yrigoyen's economic policies outside the sphere of purely partisan politics." See Galarza, p. 317.

123. Peter H. Smith, *Argentina and the Failure of Democracy*, pp. 90–97; Sánchez Sorondo, p. 77.

124. Great Britain, [1], 371/14196, May 24, 1930, Millington-Drake (Buenos Aires) to Foreign Office; Castro, pp. 11–12; Sommariva, pp. 231–33, 238.

125. Rodríguez, pp. 425–27, 447–53; Correas, p. 502; Great Britain, [1], 371/14196, May 24, 1930, Millington-Drake (Buenos Aires) to Foreign Office, pp. 32–33; Robert Richard Smith, pp. 220–34, 244–45.

126. Sommariva, p. 240; Robert Richard Smith, pp. 231–32, 256. Also see Mayo, Andino, and García Molina, pp. 150–51 and 191. These authors argue that "There is no direct relationship among the chronology of the revolutionary outbreak, the senatorial elections in Mendoza and San Juan, and the petroleum nationalization. Those who affirm this view have not taken the time to count the senators (p. 191). Their argument, however, fails to consider the possibility of additional provincial interventions, and it also fails to consider the legitimacy crisis that Yrigoyen's intervention policy created among the political opposition.

127. Examples of writers who have employed this argument include Scalabrini Ortiz, pp. 163–75; Luna, *Yrigoyen*, p. 383; del Mazo, vol. 2, pp. 138–39; and Frondizi, *Petróleo y política*, p. xlix.

128. Perette is quoted in Novau, p. 21.

129. Letter from Oscar Alende to José Novau, Sep. 1, 1958, quoted in Novau, p. 17.

130. Uriburu's first Minister of the Interior was Matías Sánchez Sorondo, a legal adviser to Standard Oil. The Minister of Public Education was Ernesto Padilla, the brother of Guillermo Padilla, whose distribution firm was a partner of WICO. Several other ministers, including Ernesto Bosch (Foreign Relations), Octavio S. Pico (Public Works), and Horacio Beccar Varela (Agriculture), were present or former employees of British oil companies. See Hollander, p. 481.

131. Johnson, "A Theoretical Model of Economic Nationalism in New and Developing States," p. 6. Mayo, Andino, and García Molina conclude (p. 191) that no hard evidence exists that the oil companies conspired with the military to carry out the 1930 coup.

Notes to Chapter 6

1. Larra, pp. 152–57, 162–66.

2. *Ibid.*, pp. 169–84; Mosconi, *El petróleo argentino*, p. 5. The battles of Junín and Ayacucho (1824) in Peru marked the final military defeat of Spain on the South American continent.

3. Mosconi, *Dichos y hechos*, p. 5.

4. Frondizi, *Petróleo y política*, pp. 329–30; Navarro Gerassi, pp. 69–70; Rumbo, pp. 83–84.

5. Hollander, pp. 506–25.

6. Gómez, pp. 11–15, 54–57; Frank, *South American Journey*, pp. 192–93.

7. The Roca-Runciman Pact and its effects on the Argentine economy long have generated intense scholarly polemics. Recent attempts to evaluate the Pact analytically include Peter H. Smith, *Politics and Beef*, pp. 140–47; Villanueva, pp. 65–66; and Tulchin, "Foreign Policy," pp. 95–98.

8. For criticism of Justo's policy, see Frondizi, *Petróleo y política*, pp. 363–65; and Kaplan, *Economía y política*, p. 24.

9. Silenzi de Stagni, *El petróleo argentino*, p. 57; Julio V. González, p. 90; Zinser, "Alternative Means of Satisfying Argentine Petroleum Demand," p. 52.

10. Hollander, pp. 554–648.

11. Larson, Knowlton, and Popple, p. 118; Frondizi, *Petróleo y política*, pp. 341–42, 377–82; Zinser, "Alternative Means of Satisfying Argentine Petroleum Demand," pp. 52–54.

12. Fanning, p. 19; Larson, Knowlton, and Popple, p. 118.

13. "Prohibición de la exportación de petróleo y fiscalización de la importación," *Boletín de Informaciones Petroleras*, 13 (Jul. 1936), pp. 10–14; Zinser, "Alternative Means of Satisfying Argentine Petroleum Demand," pp. 52–53.

14. Julio V. González, pp. 61–62, 83; Silenzi de Stagni, *El petróleo argentino*, pp. 58–60. For Justo's decrees, see "Prohibición de la exportación de petróleo y fiscalización de la importación," pp. 10–14. The American oilman is quoted in Cox (Buenos Aires) to Secretary of State, May 22, 1936, in United States, [1], p. 184.

15. Baldrich, *El problema del petróleo*, pp. 5–19; Frondizi, *Petróleo y política*, pp. 392–98; Hollander, pp. 662–64.

16. Silenzi de Stagni, *El petróleo argentino*, pp. 65–66; Frondizi, *Petróleo y política*, p. 401; Villar Araujo, 24 (Apr. 1975), p. 19. For attempts by national industry to supply YPF, see Cochran and Reina, pp. 54–55, 193–94.

17. Díaz Alejandro, p. 406; Little, p. 167; Di Tella and Zymelman, p. 81.

18. Germani, p. 75.

19. Little, pp. 162–78; Halperín Donghi; and Peter H. Smith, "The Social Base of Peronism," pp. 55–73.

20. "El homenaje tributado por SUPE al General Perón y a su señora esposa," *Boletín de Informaciones Petroleras*, 26 (Aug. 1949), p. 1. For an overview of YPF's improvements of workers' housing and labor conditions, see the chapter "Acción Social de YPF" in Argentina, [32], pp. 75–85.

21. Epstein, pp. 616–17, 629–30.

22. Díaz Alejandro, p. 446; Rock, "The Survival and Restoration of Peronism," pp. 186–90.

23. Julio V. González, p. 3.

24. Quoted in Palacios, p. 23. Also see Edwards, p. 161.

25. Perón outlined his oil policy in "La palabra del presidente de la nación," *Boletín de Informaciones Petroleras*, 25 (Jan. 1948), pp. 1–4. For the petroleum

provisions of Perón's 1947 five-year plan, see "La colaboración de YPF en el Plan Quinquenal," *ibid.*, 24 (Feb. 1947), pp. 81–88.

26. Cochran and Reina, p. 193; Grew (Department of State) to Reed (Buenos Aires), Feb. 3, 1945, in United States, [2], p. 527; Thorp (Deputy to the Assistant Secretary of State for Economic Affairs) to Paul (Special Assistant to the Secretary of Commerce), Dec. 29, 1945, in United States, [2], p. 559.

27. Kaplan, *Economía y política*, pp. 44–46.

28. Silenzi de Stagni, *El petróleo argentino*, pp. 101, 118–19, 146; Argentina, [32], pp. 7, 21; Villar Araujo, 24 (Apr. 1975), p. 20.

29. For the terms of the contract, see Palacios, pp. 22–36; and Edwards, p. 161. Perón's 1955 message to Congress may be found in a valuable collection of documents, Roggi, ed., *Argentina: Petróleo y soberanía, 1955–1964*, pp. 4/2–4/7. Also see Díaz Alejandro, p. 354.

30. For the position of the Radical National Committee, see Whitaker, pp. 104–5. Also see Silenzi de Stagni, *El petróleo argentino*, pp. 132–33; Palacios, pp. 23, 31–34; and Roggi, pp. 4/8–4/20.

31. For Perón's statement, see Perón, p. 292. Also see Rumbo, p. 199.

32. The text of Lonardi's address of Oct. 26, 1955, may be found in Whitaker, p. 162. Also see Kaplan, *Economía y política*, p. 139.

33. Whitaker, p. 112.

34. Snow, pp. 72–79; Frondizi, *Petróleo y política*, pp. lxv–lxxii.

35. Rock, "The Survival and Restoration of Peronism," pp. 194–96; Díaz Alejandro, pp. 528, 538.

36. For summaries of Frondizi's economic development plan, see Zuvekas, pp. 46–51; and Mallon and Sourrouille, p. 20. On the oil import situation, see Edwards, p. 160; and Sábato, p. 16.

37. Odell, "The Oil Industry," pp. 285–86; Sábato, pp. 18–20.

38. For details of the production contracts, see Zinser, "Alternative Means of Satisfying Argentine Petroleum Demand," pp. 180–92; and Volski, p. 257.

39. Volski, pp. 261–62.

40. Odell, "The Oil Industry," p. 287. For details on production results, see Zinser, "Alternative Means of Satisfying Argentine Petroleum Demand," pp. 57–62; and Zinser, "Alternative Means of Meeting Argentina's Petroleum Requirements," pp. 189–215. Also see Sábato, pp. 23–24; and Hartshorn, p. 229.

41. Silenzi de Stagni, "Prólogo," p. 13; Casella and Clara, p. 25; and Odell, "The Oil Industry," p. 286.

42. Hanson, "The End of the Good-Partner Policy," p. 67. For Frondizi's own defense, see his *Petróleo y nación*.

43. For SUPE's statement, see *La Prensa*, Nov. 10, 1958, p. 4.

44. Selser, pp. 71–73. *La Prensa*, Nov. 2, 1958, p. 5; Nov. 5, 1958, p. 1; Nov. 10, 1958, p. 1; Nov. 12, 1958, p. 1; Nov. 18, 1958, p. 1.

45. Rock, "The Survival and Restoration of Peronism," pp. 203–5; Díaz Alejandro, pp. 487, 528, 538.

46. Zinser, "Alternative Means of Satisfying Argentine Petroleum Demand," p. 62.

47. Snow, pp. 106–7; Edwards, pp. 165–71, 182; Tanzer, pp. 353–54. Arturo Sábato challenged Illia's arguments on economic grounds. See Sábato, p. 90.

48. Ceresole, pp. 40–42; Estep, pp. 50–51, 60; Corbett, pp. 117–18; O'Donnell, pp. 208–10.

49. Quoted in Villar Araujo, no. 25 (May 1975), p. 7. For other analyses of Onganía's oil policy, see Edwards, pp. 184–85; and Tanzer, p. 356.

50. "Stagnation in Argentina," *Petroleum Economist*, 40 (Mar. 1973), pp. 98–99; "Argentina—State Marketing Monopoly," *ibid.*, 41 (Oct. 1974), p. 389; Villar Araujo, no. 25 (May 1975), p. 7.

51. *World Petroleum Report*, 19 (1973), p. 98; "Progress Report: Argentina Looking for Energy Policy," *Latin America Economic Report*, 3 (Nov. 21, 1975), p. 182.

52. "Argentina Struggles for an Energy Breakthrough," *Latin America Economic Report*, 5 (Sep. 30, 1977), pp. 157–58.

53. Edwards, pp. 175–76; Sábato, pp. 40–41, 71; "Progress Report: Argentina Looking for Energy Policy," *Latin America Economic Report*, 3 (Nov. 21, 1975), p. 182; *Petroleum Economist*, 44 (Sep. 1977), p. 374.

54. *World Petroleum Report*, 19 (1973), p. 98; *La Opinión*, Dec. 13, 1974, p. 11; "Argentina: The Energy Sector," *Bank of London & South America Review*, 9 (Apr. 1975), pp. 196–97, 201.

55. Donald O. Croll, "Oil Nationalism Modified," *The Review of the River Plate*, 161 (Apr. 29, 1977), p. 551.

56. "YPF: Trying to Make Ends Meet," *The Review of the River Plate*, 161 (Jan. 12, 1977), p. 20; "Oil on the Move," *ibid.*, 162 (Aug. 10, 1977), pp. 194–95; Croll, "Oil Nationalism Modified," pp. 551–52.

57. "Argentina Looks Abroad for Off-shore Oil Expertise," *Latin America Economic Report*, 4 (Feb. 6, 1976), p. 23; Croll, "Oil Nationalism Modified," p. 552. Under a "risk contract," the company typically bears the full cost of unsuccessful exploration. When successful, the contractor receives all the oil until exploration costs are met, and typically 40 to 60 percent of production thereafter.

58. For references to opposition within the army, see Croll, "Oil Nationalism Modified," p. 552; and "Oil on the Move," *The Review of the River Plate*, 162 (Aug. 10, 1977), pp. 194–95.

59. For a penetrating analysis of the role of a nationalistic petroleum ideology in shaping development strategy in another Latin American country, see the important work by Tugwell, pp. 151–53.

60. For Mosconi's fascination with the mixed-company model of Anglo-Persian, see Palacios, p. 100.

61. For an analysis of the appeal of economic nationalism among the Peronist masses, see Kirkpatrick, pp. 183–86.

62. For background on ANCAP, see Hanson, *Utopia in Uruguay*, pp. 93–98. Also see Mosconi, *El petróleo argentino*, pp. 231–33; Pérez Prins, pp. 15–16; and Odell, "Oil and State in Latin America," p. 662.

63. Pérez Prins, pp. 16–17. Mosconi's speech is in *Dichos y hechos*, p. 224.

64. For Froianini's remarks, see "YPF Bolivianos," *Boletín de Informaciones Petroleras*, 14 (Jan. 1937), p. 118. For a thorough review of the long-simmering dispute between Standard and the Bolivian government that preceded the expropriation, see Klein, pp. 47–63.

65. Frondizi, *Petróleo y política*, p. 373; Ostría Gutiérrez, pp. 267–81; Klein, pp. 47, 66, 71–72; Wood, pp. 170, 177, 187–88.

66. Wirth, pp. 136–53; Peter Seaborn Smith, pp. 35–38.

67. Mosconi, *Dichos y hechos*, pp. 225–26.

68. Wirth, pp. 151–71; Peter Seaborn Smith, pp. 38–55; Penna Marinho, pp. 350–54, 361. For Horta's visit to Comodoro Rivadavia, see "Grata visita: autoridades brasileñas en Comodoro Rivadavia," *Boletín de Informaciones Petroleras*, 16 (May 1939), pp. 68–69. For his 1947 speech, see Horta Barbosa, pp. 49–62.

69. Wirth, pp. 184–94, 212–13.

70. "Brazil—Oil Imports a Heavy Burden," *Petroleum Economist*, 41 (Mar. 1974), p. 110; "Brazil's Oil Outlook Still Obstinately Cloudy," *Latin America Economic Report*, 3 (Sep. 19, 1975), p. 147; "Piecemeal Revelation of Brazilian Oil Contracts," *Latin America Economic Report*, 4 (Feb. 13, 1976), p. 26.

71. "Una crisis que favorece el monopolio," *Boletín del Petróleo* (México), 30 (Nov.-Dec. 1930), p. 246. Also see Meyer, pp. 297–99.

72. Larra, pp. 64–65. For the background to Mexico's 1938 expropriations, see Powell, pp. 22–36; Robert Freeman Smith, pp. 77–255; and Meyer, pp. 83–346.

73. Penna Marinho, p. 199.

74. "Latin American Prospects," *Petroleum Economist*, 41 (Jan. 1974), pp. 25–28; "OLADE Still Unable to Go Beyond Talking Stage," *Latin America Economic Report*, 3 (Sep. 19, 1975), p. 145.

Bibliography

Research in twentieth-century Argentine history is often a frustrating experience, for government archives generally are inaccessible and prominent families guard their papers from the historian's view. Problems of access to materials become particularly acute when the research touches on a sensitive area, which the petroleum industry certainly is. The papers of such key figures as Enrique Mosconi and Hipólito Yrigoyen are unobtainable or have been destroyed, and the archives of YPF and the private oil companies are unavailable. YPF once maintained a large archive open to the public, but the first Perón government, for reasons never explained, moved it to a suburban warehouse, where those materials not lost now languish decaying and inaccessible. Similarly, in reply to a request for information, Mr. J. M. Freyman of the Exxon Corporation, the successor to Jersey Standard, informed me that records of Argentine operations in the 1920's and 1930's could not be located (letter of Sep. 5, 1975).

Despite these obstacles, a great deal of source material does exist for the study of Argentine petroleum history. The published *Memorias* of YPF, as well as those of its predecessor company and of other government agencies, proved highly useful. Congressional debates were an important source for information on petroleum politics, as was the rich and varied commentary of the Argentine press. U.S. and British diplomatic correspondence contained much information on the foreign oil companies, YPF, and Argentine politics. And of course there were the hundreds of books, pamphlets, and articles published about Argentine and world petroleum affairs throughout this century; they yielded much of the evidence on which this book's analysis is based.

A word about the organization of this Bibliography. Official and semiofficial publications of Argentina, Great Britain, and the United States are listed under the respective country and are preceded by bracketed numbers for ease of reference in the Notes. I have chosen not to list separately here the many periodicals and newspapers I consulted during my research; full citations are given when necessary in the Notes.

Alsacio Cortázar, Miguel Mario. *Burguesía argentina y petróleo nacional*. Buenos Aires: Avanzar, 1969.

American Petroleum Institute, Division of Public Relations. *Petroleum Facts and Figures*. Baltimore, Md.: American Petroleum Institute, 1929.

Argentina, Official and Semiofficial Publications:
[1] Academia Nacional de la Historia. *Historia argentina contemporánea, 1862– 1930*. 4 vols. Buenos Aires: "El Ateneo," 1965–67.
[2] Cámara de Diputados de la Nación. *Diario de sesiones de la Cámara de Diputados*. Buenos Aires, 1908–30.
[3] ———. *Informe de la Comisión Investigadora de los Trusts. Septiembre de 1919*. Buenos Aires, 1919.
[4] ———. *Yacimientos Petrolíferos Fiscales: Antecedentes para su explotación. Iniciativas parlamentarias*. 3 vols. Buenos Aires, 1924.
[5] Cámara de Senadores de la Nación. *Diario de sesiones de la Cámara de Senadores*. Buenos Aires, 1909–29.
[6] *Código de minería de la República Argentina*. Buenos Aires, 1937.
[7] *Código de minería de la República Argentina sancionado por ley del Honorable Congreso de 8 de diciembre de 1886*. Buenos Aires, 1887.
[8] Dirección General de Estadística de la Nación. *Anuario del comercio exterior de la República Argentina, año 1930*. Buenos Aires, 1931.
[9] Dirección General de Explotación del Petróleo de Comodoro Rivadavia. *Memoria correspondiente a los años 1912/1913 presentada a S.E. el Señor Ministro de Agricultura de la Nación*. Buenos Aires, 1914.
[10] ———. *(Nota fundando su pedido de 2.000.000 $ m/n para proseguir los trabajos)*. Buenos Aires, 1911.
[11] ———. *El petróleo de Comodoro Rivadavia: Informe de la Dirección General a S.E. el Señor Ministro de Agricultura, fundando un programa de trabajos y el presupuesto de gastos para los años 1913 y 1919*. Buenos Aires, 1912.
[12] Dirección General de Yacimientos Petrolíferos Fiscales. *Antecedentes y desarrollo de la industria del petróleo en la República Argentina*. Buenos Aires[?], 1931[?].
[13] Explotación Nacional del Petróleo de Comodoro Rivadavia. *Memoria de la Explotación Nacional del Petróleo de Comodoro Rivadavia correspondiente al año 1918*. Buenos Aires, 1919.
[14] ———. *Memoria de la Explotación Nacional del Petróleo de Comodoro Rivadavia correspondiente al año 1919*. Buenos Aires, 1920.
[15] ———. *Memoria de la Explotación Nacional del Petróleo de Comodoro Rivadavia correspondiente al año 1920*. Buenos Aires, 1921.
[16] *Las grandes defraudaciones: West India Oil Company (WICO), Compañía Nacional de Petróleos: informaciones sobre el proceso judicial: demostración de fraude*. Buenos Aires, 1923.
[17] Ministerio de Agricultura. *Anales del Ministerio de Agricultura, Sección Geologia, Mineralogia y Mineria* (Aug. 1904), Buenos Aires, 1904.
[18] Ministerio de Agricultura de la Nación. Dirección General de Yacimientos Petrolíferos Fiscales. *Memoria de la Dirección General de Yacimientos Petrolíferos Fiscales correspondiente al año 1922*. Buenos Aires, 1924.

[19] ———. ———. *Memoria de la Dirección General de Yacimientos Petrolíferos Fiscales correspondiente al año 1925*. Buenos Aires, 1926.

[20] ———. ———. *Memoria de la Dirección General de Yacimientos Petrolíferos Fiscales correspondiente al año 1927*. Buenos Aires, 1928.

[21] ———. ———. *Memoria de la Dirección General de Yacimientos Petrolíferos Fiscales correspondiente al año 1928*. Buenos Aires, 1929.

[22] ———. ———. *Memoria de la Dirección General de Yacimientos Petrolíferos Fiscales, año 1929*. Buenos Aires, 1930.

[23] ———. ———. *Memoria de la Explotación del Petróleo de Comodoro Rivadavia correspondiente al año 1921*. Buenos Aires, 1923.

[24] ———. ———. *Presupuesto general del año 1924 y plan financiero para los años 1924–1927*. Buenos Aires, 1924.

[25] ———. ———. *Reglamento orgánico de la Dirección General de Yacimientos Petrolíferos Fiscales. Reglamento interno de la Comisión Administrativa de los Yacimientos Petrolíferos Fiscales*. Buenos Aires, 1928.

[26] ———. Yacimientos Petrolíferos Fiscales. *Recopilación de leyes, decretos y resoluciones sobre materia petrolífera (1907–1933)*. Buenos Aires, 1938.

[27] Ministerio de Guerra. *Memoria del Departamento de Guerra presentada al Honorable Congreso Nacional correspondiente al año 1924–1925*. Buenos Aires, 1925.

[28] ———. *Memoria del Departamento de Guerra presentada al Honorable Congreso Nacional correspondiente al año 1926–1927*. Buenos Aires, 1927.

[29] Ministerio del Interior. *Memoria del Ministerio del Interior presentada al Honorable Congreso de la Nación, 1920–1921*. Buenos Aires, 1921.

[30] Ministerio de Marina. *Memoria del Ministerio de Marina correspondiente al ejercicio 1916–1917*. Buenos Aires, 1917.

[31] ———. *Memoria del Ministerio de Marina correspondiente al ejercicio 1917–1918*. Buenos Aires, 1918.

[32] Secretaría de Industria y Comercio de la Nación. Yacimientos Petrolíferos Fiscales. *Memoria correspondiente al año 1947*. Buenos Aires, 1948.

[33] Superintendente del Censo. *Primer censo de la República Argentina verificado en los días 15, 16 y 17 de septiembre de 1869*. Buenos Aires, 1872.

[34] Yacimientos Petrolíferos Fiscales. *Desarrollo de la industria petrolífera fiscal, 1907–1932*. Buenos Aires, 1932.

Bain, H. Foster, and Thomas T. Read. *Ores and Industry in South America*. New York: Council on Foreign Relations, 1934.

Baldrich, Alonso. *El petróleo: Su importancia comercial, industrial y militar. Legislación petrolera*. Buenos Aires: "La Argentina" 1927.

———. *El problema del petróleo y la guerra del Chaco*. Buenos Aires: "Revista Americana de Buenos Aires," 1934.

Barcía Trelles, Camilo. *El imperialismo del petróleo y la paz mundial*. Valladolid: Universidad de Valladolid, Sección de Estudios Americanistas, 1925.

Barrera, Raúl. *El petróleo de Comodoro Rivadavia: Contribución al estudio financiero de los yacimientos fiscales*. Buenos Aires: Instituto Geográfico Militar, 1919.

Bayá, Rómulo. *"Yo acuso."* Buenos Aires: Tor, 1926.

Bayer, Osvaldo, *Los vengadores de la Patagonia trágica.* 2 vols. Buenos Aires: Galerna, 1972.

Bazán, Pedro. *El fomento económico de La Rioja.* Buenos Aires: Prats, 1927.

Benham, Frederic. *Great Britain Under Protection.* New York: Macmillan, 1941.

Bianco, José. *La crisis: Nacionalización del capital extranjero.* Buenos Aires: Mendesky—Augusto Sabourín é Hijo 1916.

Bonardi, Silvio E. "La cuestión petrolífera en Salta," *Revista de Ciencias Económicas,* 31 (Nov. 1928), pp. 2539–50, 31 (Dec. 1928), pp. 2630–44, and 32 (Jan. 1929), pp. 51–68.

Braden, Spruille. *Diplomats and Demagogues: The Memoirs of Spruille Braden.* New Rochelle, N.Y.: Arlington House, 1971.

Bradley, J. B. "Argentine Fuel and Power and the International Coal Trade," *Comments on Argentine Trade,* 9 (May 1930), pp. 40–48.

Brady, George S. *Argentine Petroleum Industry and Trade.* Washington D.C.; Dep't of Commerce, Bureau of Foreign and Domestic Commerce, 1923.

———. *Railways of South America. Part I: Argentina.* Washington, D.C.: Dep't of Commerce, Bureau of Foreign and Domestic Commerce, 1926.

Buchanan, James E. "Politics and Petroleum Development in Argentina, 1916–1930." Ph.D. diss., Univ. of Mass., 1973.

Bucich Escobar, Ismael. *Historia de los presidentes argentinos.* Buenos Aires: Roldán, 1934.

Bunge, Alejandro E. *The Cost of Living in the Argentine Republic: Wages and Output.* Buenos Aires: no pub., 1928.

———. *Ferrocarriles argentinos: Contribución al estudio del patrimonio nacional.* Buenos Aires: Mercatali, 1918.

———. *Una nueva Argentina.* Buenos Aires: Kraft, 1940.

———. "La legislación del petróleo," *Revista de Economía Argentina,* 21 (Oct. 1928), p. 289.

———. "El problema económico del petróleo," *Revista de Economía Argentina,* 24 (June 1930), pp. 401–36.

Bunge, Alejandro E., and Miguel A. Sasot, eds. *El estado industrial y comerciante.* Buenos Aires: "Economía Argentina," 1932.

Bunge, Augusto. *La guerra del petróleo en la Argentina.* Buenos Aires: no pub., 1935.

Caballero, Ricardo. *Yrigoyen: La conspiración civil y militar del 4 de febreo de 1905.* Buenos Aires: Raigal, 1951.

Cáceres Cano, Severo G. *Apuntes para la historia de un esfuerzo argentino.* Buenos Aires: Theoría, 1972.

Cantón, Darío. *Materiales para el estudio de la sociología política en la Argentina.* 2 vols. Buenos Aires: Centro de Investigaciones Sociales, Instituto Torcuato di Tella, 1968.

Cárcano, Miguel Angel. "Ensayo histórico sobre la presidencia de Roque Sáenz Peña," in Argentina [1], vol. 1, part 2, pp. 135–91.

Caro Figueroa, Gregorio. *Historia de la gente decente en el norte argentino.* Buenos Aires: Mar Dulce, 1970.

Cartavio, A. R. "Datos sobre algunas industrias argentinas." Ms. in the Biblioteca Tornquist, Buenos Aires; 1918.

Casal, Horacio N. *El petróleo*. Buenos Aires: Centro Editor de América Latina, 1972.

Casella, Alberto T., and Alejandro Clara. *Petróleo, soberanía y paz*. Buenos Aires: Platina, 1963.

Castro, Isaac E. *El irigoyenismo contra la organización nacional: Las intervenciones federales desde 1916 hasta 1929*. Buenos Aires, no pub., 1929.

Centro de Estudios General Mosconi. *Los tratantes de petróleo. Tomo I. Los hechos (1955–1962)*. Buenos Aires: Achával Solo, 1973.

Ceresole, Norberto. *El ejército y la crisis política argentina*. Buenos Aires: Política Internacional, 1970.

Chavarría, Juan Manuel. *El problema económico y social de Catamarca*. Buenos Aires: "El Ateneo," 1926.

Coca, Joaquín. *El contubernio: Selección*. Buenos Aires: Coyocán, 1961.

Cochran, Thomas C., and Ruben E. Reina. *Entrepreneurship in Argentine Culture: Torcuato di Tella and S. I. A. M*. Philadelphia: Univ. of Pennsylvania Press, 1962.

Colombo, Luis. *Levántate y anda*. Buenos Aires: M. Gleizer, 1929.

——. *El petróleo argentino y la necesidad de su legislación*. Buenos Aires: Unión Industrial Argentina, 1927.

Comité Universitario Radical, Junta Central. *El petróleo argentino. Ciclo de conferencias en pro de su nacionalización y explotación por el estado*. Buenos Aires: "Capano," 1930.

Compañía Nacional de Petróleos Ltda. *El gravamen aduanero del kerosene*. Buenos Aires: no pub., 1916.

Confederación Argentina del Comercio, de la Industria y de la Producción. *Actas de la Tercera Conferencia Económica Nacional (2–12 de julio 1928)*. Buenos Aires: Compañía Impresora Argentina, 1928.

Corbett, Charles C. *The Latin American Military as a Socio-Political Force: Case Studies of Bolivia and Argentina*. Coral Gables, Fla.: Univ. of Miami Center for Advanced International Studies, 1972.

Cornejo, Atilio. "La cuestión de petróleo," *Revista Argentina de Ciencias Políticas*, 33 (Oct. 1926), pp. 96–121.

Correa, Guillermo. "Una cuestión del momento," *Revista de Economía Argentina*, 19 (Sep.-Oct. 1927), pp. 259–64.

Correas, Edmundo. "Historia de Mendoza (1862–1930)," in Argentina [1], vol. 4, part 1, pp. 469–502.

Craviotto, José A. "La minería y el petróleo (1862–1930)," in Argentina, [1], vol. 3, pp. 463–573.

Croll, Donald O. "Oil Nationalism Modified," *The Review of the River Plate*, 161 (Apr. 29, 1977), pp. 551–52.

Davel, Ricardo J. "Petróleos argentinos," in Universidad Nacional de Buenos Aires, Facultad de Ciencias Económicas, *Investigaciones de Seminario*, 1 (1917), pp. 147–222.

Davenport, E. H., and Sidney Russell Cooke. *The Oil Trusts and Anglo-American Relations*. New York: Macmillan, 1924.

Denny, Ludwell. *We Fight for Oil*. New York: Knopf, 1928.

De Novo, John A. "The Movement for an Aggressive American Oil Policy Abroad, 1918–1920," *The American Historical Review*, 61 (July 1956), pp. 854–76.

Deterding, Henri. *An International Oilman*. London: Harper and Brothers, 1934.

Díaz Alejandro, Carlos F. *Essays on the Economic History of the Argentine Republic*. New Haven, Conn.: Yale Univ. Press, 1970.

Dickmann, Adolfo, Héctor Iñigo Carrera, and Adolfo Rubenstein. *En defensa del petróleo nacional y por la dignidad de la función pública*. Buenos Aires: La Vanguardia, 1932.

Di Tella, Guido, and Manuel Zymelman. *Las etapas del desarrollo económico argentino*. Buenos Aires: Editorial Universitaria de Buenos Aires, 1967.

Di Tella, Torcuato S., ed. *Argentina, sociedad de masas*. Buenos Aires: Editorial Univeristaria de Buenos Aires, 1965.

Dorfman, Adolfo. *Evolución industrial argentina*. Buenos Aires: Losada, 1942.

Dreier, Katherine S. *Five Months in the Argentine: From a Woman's Point of View*. New York: Frederic Fairchild Sherman, 1920.

Duhau, Luis. *Los elevadores de granos en el Canadá*. Buenos Aires: Sociedad Rural Argentina, 1928.

———. "Fomento de las relaciones económicas entre Estados Unidos y la Argentina." Mimeo. Washington, D.C.: Unión Panamericana, 1927.

———. "Intercambio comercial con Estados Unidos," *Revista de Ciencias Económicas*, 29 (Sept. 1927), pp. 995–1008.

Duvall, Laurel. "The Production and Handling of Grain in Argentina." *Yearbook of the Department of Agriculture, 1915*. Washington, D.C.: Dep't of Agriculture, 1916.

Dye, Alexander V. "Commerce Between Argentina and the United States in 1927," *Comments on Argentine Trade*, 7 (Jan. 1928), p. 21.

Easum, Donald Boyd. "The British-Argentine-United States Triangle: A Case Study in International Relations." Ph.D. diss., Princeton Univ., 1953.

Edwards, Gertrude G. "The Frondizi Contracts and Petroleum Self-Sufficiency in Argentina," in Mikesell, ed., *Foreign Investment in the Petroleum and Mineral Industries*, pp. 157–88.

Epstein, Edward C. "Politicization and Income Redistribution in Argentina: The Case of the Peronist Worker," *Economic Development and Cultural Change*, 23 (Jul. 1975), pp. 615–32.

Estep, Raymond. *The Argentine Armed Forces and Government*. Maxwell Air Force Base, Ala.: Aerospace Studies Institute, Air University, 1970.

Etchepareborda, Roberto. "Notas sobre la presidencia de Marcelo T. de Alvear," *Revista de Administración Militar y Logística*, 459 (Mar. 1976), pp. 115–51.

———. "La segunda presidencia de Hipólito Yrigoyen y la crisis de 1930," in Argentina [1], vol. 1, part 2, pp. 347–75.

Falcoff, Mark. "Intellectual Currents," in Falcoff and Dolkart, eds., *Prologue to Perón*, pp. 110–35.

Falcoff, Mark, and Ronald H. Dolkart, eds. *Prologue to Perón: Argentina in Depression and War, 1930–1943*. Berkeley; Univ. of California Press, 1975.

Fanning, Leonard M. *American Oil Operations Abroad*. New York: McGraw-Hill, 1947.

Feis, Herbert. *Petroleum and American Foreign Policy*. Stanford, Calif.: Food Research Institute, 1944.

Fennell, Lee Cameron. "Class and Region in Argentina: A Study of Political Cleavage, 1916–1966." Ph.D. diss., Univ. of Florida, 1970.

Ferrer, James. "United States-Argentine Economic Relations, 1900–1930." Ph.D. diss., Univ. of Calif., Berkeley, 1964.

Ferns, H. S. *Argentina*. New York: Praeger, 1969.

Fliess, Felipe. *El petróleo y Comodoro Rivadavia*. Buenos Aires: Pedro Preusche, 1922.

Fodor, Jorge G., and Arturo A. O'Connell. "La Argentina y la economía atlántica en la primera mitad del siglo XX," *Desarrollo Económico*, 13 (Apr.-Jun. 1973), pp. 3–65.

Ford, A. G. "British Investment and Argentine Economic Development, 1880–1914," in Rock, ed., *Argentina in the Twentieth Century*, pp. 12–40.

———. *The Gold Standard, 1880–1914: Britain and Argentina*. Oxford: Clarendon Press, 1962.

Frank, Waldo. *America Hispana: A Portrait and a Prospect*. New York: Charles Scribner's Sons, 1931.

———. *South American Journey*. New York: Duell, Sloan and Pearce, 1943.

Frankel, P. H. *Mattei: Oil and Power Politics*. New York: Praeger, 1966.

Fraser, John F. *The Amazing Argentine*. New York: Funk & Wagnells, 1914.

Frondizi, Arturo. *Petróleo y nación*. Buenos Aires: Transición, 1963.

———. *Petróleo y política*. 2d ed. Buenos Aires: Raigal, 1955.

Galarza, Ernest. "Argentina's Revolution and its Aftermath," *Foreign Policy Reports*, 7 (Oct. 28, 1931), pp. 309–22.

Galletti, Alfredo. *La política y los partidos*. Buenos Aires: Fondo de Cultura Económica, 1961.

Gallo, Ezequiel (h.), and Silvia Sigal. "La formación de los partidos políticos contemporáneos: La U.C.R. (1890–1916)," in Di Tella, ed., *Argentina, sociedad de masas*, pp. 124–76.

Gálvez, Manuel. *Vida de Hipólito Yrigoyen*. 2d ed. Buenos Aires: Kraft, 1939.

Germani, Gino. *Estructura social de la Argentina: Análisis estadístico*. Buenos Aires: Raigal, 1955.

Gibb, George Sweet, and Evelyn H. Knowlton. *The Resurgent Years, 1911–1927*. New York: Harper & Brothers, 1956. [This is part of a series entitled *History of the Standard Oil Company (New Jersey)* and prepared under the auspices of the Business History Foundation.]

Goldwert, Marvin. "The Argentine Revolution of 1930: The Rise of Modern Militarism and Ultra-Nationalism in Argentina." Ph.D. diss., Univ. of Texas, 1962.

Gómez, Rufino. *La gran huelga petrolera de Comodoro Rivadavia (1931–1932) en el recuerdo del militante obrero y comunista Rufino Gómez*. Buenos Aires: Ediciones Centro de Estudios, 1973.

González, Juan B. *El encarecimiento de la vida en la República Argentina*. Buenos Aires: Las Ciencias, 1908.

González, Julio V. *Nacionalización del petróleo*. Buenos Aires: "El Ateneo," 1947.

Goodwin, Paul B. *Los ferrocarriles británicos y la U.C.R. (1916–1930)*. Buenos Aires: La Bastilla, 1974.

————. "The Politics of Rate-Making: The British-owned Railways and the Unión Cívica Radical," *Journal of Latin American Studies*, 6 (Nov. 1974), pp. 257–87.

Gravil, Roger. "Anglo-U.S. Trade Rivalry in Argentina and the D'Abernon Mission of 1929," in Rock, ed., *Argentina in the Twentieth Century*, pp. 41–65.

Great Britain, Official and Semiofficial Publications:

[1] Foreign Office. *General Correspondence: Political, 1920–30 (Series F.O. 371)*. Ms. in the Public Record Office, London.

[2] House of Commons. *Parliamentary Debates*. 1929.

[3] ————. *Sessional Papers*. "Correspondence Between His Majesty's Government and the United States Ambassador Respecting Economic Rights in Mandated Territories. Presented to Parliament by Command of His Majesty." 1921 (Cmd. 1226).

[4] ————. ————. "Despatch to His Majesty's Ambassador at Washington Enclosing a Memorandum on the Petroleum Situation. Presented to Parliament by Command of His Majesty." 1921 (Cmd. 1351).

[5] ————. ————. "Report on the Economic and Industrial Situation of the Argentine Republic for the Year 1919 by Mr. H. O. Chalkley, Commercial Secretary to H. M. Legation, Buenos Aires. Presented to Parliament by Command of His Majesty." 1921 (Cmd. 895).

Guevara, Salvador. "Favorecer a la industria privada." *Revista Militar*, 26 (May 1926), pp. 737–39.

————. "Industria siderúgica," *Revista Militar*, 26 (Mar. 1926), pp. 397–403.

Guevara Labal, Carlos. *El General Ingeniero Enrique Mosconi: Una vida consagrada a la patria*. Buenos Aires: A. Riera, 1941.

————. *El petróleo y sus derivados en la estadística*. Buenos Aires: Ferrari Hnos., 1932.

Halperín Donghi, Tulio. "Algunas observaciones sobre Germani, el surgimiento del peronismo y los migrantes internos," *Desarrollo Económico*, 14 (Jan.-Mar. 1975), pp. 765–81.

Hanighen, Frank C. *The Secret War*. New York: John Day, 1934.

Hanson, Simon G. "The End of the Good-Partner Policy," *Inter-American Economic Affairs*, 14 (Summer 1960), pp. 63–92.

————. *Utopia in Uruguay: Chapters in the Economic History of Uruguay*. New York: Oxford Univ. Press, 1938.

Hartshorn, J. E. *Politics and World Oil Economics*. New York: Praeger, 1962.

Hassmann, Heinrich. *Oil in the Soviet Union*. Trans. by Alfred M. Leeston. Princeton, N.J.: Princeton Univ. Press, 1953.

Hermitte, Enrique. "Carbón, petróleo y agua en la República Argentina," in *Argentina*, [17], vol. 1, pp. 71–172.

———. *Las investigaciones geológicas, mineralógicas e hidrológicas en la República Argentina. Necesidad de fomentarlas*. Buenos Aires: Oficina Metereológica Argentina, 1910.

Hileman, Guillermo, *Sobre legislación del petróleo en la República Argentina*. Buenos Aires: "La Aurora," 1927.

———. "Los yacimientos petrolíferos de Cacheuta, Provincia de Mendoza," *Revista Minera*, 1 (Sep. 1929), pp. 68–76.

Hipwell, H. Hallam. "Trade Rivalries in Argentina," *Foreign Affairs*, 8 (Oct. 1929), pp. 150–54.

Hobsbawm, E. J. *Industry and Empire*. New York: Pantheon, 1968.

Hogan, Michael J. "Informal Entente: Public Policy and Private Management in Anglo-American Petroleum Affairs, 1918–1924," *Business History Review*, 48 (Summer 1974), pp. 187–205.

Hollander, Frederick Alexander. "Oligarchy and the Politics of Petroleum in Argentina: The Case of the Salta Oligarchy and Standard Oil, 1918–1933." Ph.D. diss., Univ. of Calif., Los Angeles, 1976.

Holm, Gert T. *The Argentine Grain Grower's Grievances*. Buenos Aires: Rugeroni, 1919.

Horta Barbosa, Julio Caetano. "Nuestro petróleo debe ser explotado por el gobierno," *Boletín de Informaciones Petroleras*, 25 (Jan. 1948), pp. 49–62.

Ibarguren, Carlos. *La historia que he vivido*. Buenos Aires: Peuser, 1955.

Imaz, José Luis de. "Alejandro E. Bunge, economista y sociólogo," *Desarrollo Económico*, 14 (Oct.-Dec. 1974), pp. 545–67.

———. *Los que mandan (Those Who Rule)*. Trans. by Carlos A. Astiz. Albany: State Univ. of New York Press, 1970.

Instituto Argentino del Petróleo. *El petróleo en la República Argentina*. Buenos Aires: Instituto Argentino del Petróleo, 1966.

International Petroleum Encyclopedia. Tulsa, Okla: Petroleum Publishing Co., 1972, 1975.

Jaime, Euclides E. "La personalidad moral y política del Doctor Adolfo Güemes," *Boletín de Informaciones Petrolíferas*, 7 (Feb. 1930), pp. 157–61.

Jane, Fred T., ed. *Fighting Ships (1914)*. London: Sampson, Low, Marston & Co., 1914.

Johnson, Harry G. "The Ideology of Economic Policy in the New States," in Johnson, ed., *Economic Nationalism in Old and New States*, pp. 124–41.

———. "A Theoretical Model of Economic Nationalism in New and Developing States," in Johnson, ed., *Economic Nationalism in Old and New States*, pp. 1–16.

Johnson, Harry G., ed. *Economic Nationalism in Old and New States*. Chicago: Univ. of Chicago Press, 1967.

Kaplan, Marcos. *Economía y política del petróleo argentino (1939/1956)*. Buenos Aires: Praxis, 1957.

———. *Petróleo, estado y empresas en Argentina*. Caracas: Síntesis Dosmil, 1972.

———. "La primera fase de la política petrolera argentina (1907–1916)." *Desarrollo Económico*, 13 (Jan.-Mar. 1974), pp. 775–810.

Kirkpatrick, Jeane. *Leader and Vanguard in Mass Society: A Study of Peronist Argentina*. Cambridge, Mass.: M.I.T. Press, 1971.

Klein, Herbert S. "American Oil Companies in Latin America: The Bolivian Experience," *Inter-American Economic Affairs*, 18 (Autumn 1964), pp. 47–72.

Krause, Emma, and Amalia Krause. *Corazonada de argentinos: Prehistoria del hallazgo de petróleo en Comodoro Rivadavia*. Buenos Aires: no pub., 1960.

Kurtz, Roberto. *La Argentina ante Estados Unidos*. Buenos Aires: Luis Veggia, 1928.

Lagos, M. J. *El petróleo en América: Estados Unidos y República Argentina*. Buenos Aires: L. J. Rosso, 1924.

———. *La política del petróleo: Contribución al estudio*. Buenos Aires: L. J. Rosso, 1922.

Landa, José. *Hipólito Yrigoyen visto por uno de sus médicos*. Buenos Aires: Macland S.R.L., 1958.

Larra, Raúl. *Mosconi: General del petróleo*. Buenos Aires: Futuro, 1957.

Larson, Henrietta M., Evelyn H. Knowlton, and Charles S. Popple. *New Horizons, 1927–1950*. New York: Harper & Row, 1971. [This is part of a series entitled *History of the Standard Oil Company (New Jersey)* and prepared under the auspices of the Business History Foundation.]

Lencinas, José Hipólito. *El petróleo y los jerarcas del centralismo porteño*. Mendoza, Argentina: no pub., 1958.

Lewis, W. Arthur. *Economic Survey, 1919–1939*. London: George Allen & Unwin, 1949.

Lieuwen, Edwin. *Petroleum in Venezuela: A History*. New York: Russell & Russell, 1967.

Little, Walter. "The Popular Origins of Peronism," in Rock, ed., *Argentina in the Twentieth Century*, pp. 162–78.

Lowenthal, Abraham F., ed. *Armies and Politics in Latin America*. New York: Holmes & Meier, 1976.

Lugones, Leopoldo. *La grande Argentina*. 2d ed. Buenos Aires: Huemul, 1962.

Luna, Felix. *Alvear*. Buenos Aires: Libros Argentinos, 1958.

———. *Yrigoyen*. Buenos Aires: Desarrollo, 1964.

Machado, Manuel A. *Aftosa: A Historical Study of Foot-and-Mouth Disease and Inter-American Relations*. Albany: State Univ. of New York Press, 1969.

Maizels, Alfred. *Industrial Growth and World Trade*. Cambridge, Eng.: Cambridge Univ. Press, 1963.

Mallon, Richard D., and Juan V. Sourrouille, *Economic Policymaking in a Conflict Society: The Argentine Case*. Cambridge, Mass.: Harvard Univ. Press, 1975.

Mármora, Lelio. *Migración al sur (argentinos y chilenos en Comodoro Rivadavia)*. Buenos Aires: Libera, 1968.

Marotta, Sebastián. *El movimiento sindical argentino: Su génesis y desarrollo*. 3 vols. Buenos Aires: Lacio, 1960–70.

Martínez, Alberto A. "Foreign Capital Investments in Argentina," *The Review of the River Plate*, 49 (Jun. 7, 1918), p. 1339.

Martini, Angel Romualdo. *El pocero Fuchs*. Buenos Aires: Freeland, 1959.

Mayo, Carlos A., Osvaldo R. Andino, and Fernando García Molina. *Diplomacia, política y petróleo en la Argentina (1927–1930)*. Buenos Aires: Rincón, 1976.

Mazo, Gabriel del. *El radicalismo: Ensayo sobre su historia y doctrina*. 3 vols. Buenos Aires: Guré, 1957.

McCrea, Roswell C., Thurman W. Van Metre, and George Jackson Eder. "International Competition in the Trade of Argentina," *International Conciliation*, 271 (Jun. 1931), pp. 321–487.

Melo, Carlos R. "Presidencia de José Figueroa Alcorta," in Argentina, [1], vol. 1, part 2, pp. 101–33.

Méndez, José Manuel. *Petróleos argentinos*. Buenos Aires: Universidad Nacional de Buenos Aires, Facultad de Ciencias Económicas, 1916.

Merkx, Gilbert Wilson. "Political and Economic Change in Argentina from 1870 to 1966." Ph.D. diss., Yale Univ., 1968.

Meyer, Lorenzo. *México y los Estados Unidos en el conflicto petrolero (1917–1942)*. 2d ed. México: El Colegio de México, 1972.

Miatello, Hugo (h.). *La agricultura en la Patagonia: Zona de Comodoro Rivadavia*. Buenos Aires: Ministerio de Agricultura, 1921.

Mikesell, Raymond F., et al. *Foreign Investment in the Petroleum and Mineral Industries: Case Studies of Investor-Host Country Relations*. Baltimore, Md.: Johns Hopkins Press, 1971.

Molina, Raúl A. "Presidencia de Marcelo T. De Alvear," in Argentina, [1], vol. 1, part 2, pp. 271–345.

Montenegro, Lauro. "Algunas ideas sobre la preparación integral de la nación para la guerra," *Revista Militar*, 26 (Apr. 1927), pp. 439–65; 27 (July 1927), pp. 837–54.

Mosconi, Enrique. *La batalla del petróleo: YPF y las empresas extranjeras*. Ed. Gregorio Selser. Buenos Aires: Problemas Nacionales, 1957.

———. *Creación de la 5ª arma y las rutas aéreas argentinas*. Buenos Aires: Junta Argentina de Aviación, 1941.

———. *Dichos y hechos, 1904–1938*. Buenos Aires: "El Ateneo," 1938.

———. *El petróleo argentino, 1922–1930, y la ruptura de los trusts petrolíferos inglés y norteamericano el 1° de agosto de 1929*. Buenos Aires: "El Ateneo," 1936.

———. "Prólogo," in *El petróleo del norte argentino*, pp. vii–xxxii.

Moyano Llerena, Carlos. "Alejandro E. Bunge y la independencia económica nacional," *Revista de Economía Argentina*, 50 (Apr.-Jun. 1951), pp. 37–42.

Naón, Rómulo S. *Inviolabilidad de la propiedad minera*. Buenos Aires: Muro, 1928.

Navarro Gerassi, Marysa. *Los nacionalistas*. Trans. by Alberto Ciria. Buenos Aires: Jorge Alvarez, 1968.

Nelson, John H. "Agitation for Nationalization of Oil Renewed in Argentina," *The Oil and Gas Journal*, 28 (Feb. 27, 1930), p. 90.

Novau, José. *La revolución de 1930 y el dominio del petróleo*. Rosario, Argentina: Ruiz, 1968.

O'Connor, Richard. *The Oil Barons: Men of Greed and Grandeur*. Boston: Little, Brown, 1971.

Oddone, Jacinto. *La burguesía terrateniente argentina*. 3d ed. Buenos Aires: Ediciones Populares Argentinas, 1956.

Odell, Peter R. "Oil and State in Latin America," *International Affairs*, 40 (Oct. 1964), pp. 659–73.

————. "The Oil Industry in Latin America," in Penrose, *The Large International Firm in Developing Countries*, pp. 274–300.

O'Donnell, Guillermo A. "Modernization and Military Coups: Theory, Comparisons, and the Argentine Case," in Lowenthal, ed., *Armies and Politics in Latin America*, pp. 197–243.

Olguín, Dardo. *Lencinas: El caudillo radical. Historia y mito*. Mendoza, Argentina: Vendimiador, 1961.

Oneto, Ricardo. *El petróleo argentino y la soberanía nacional*. Buenos Aires: Ferrari Hnos., 1929.

Orzábal Quintana, Arturo. "Los soviets y el petróleo del Caucaso," *Nosotros*, 22 (Nov. 1928), pp. 162–87.

Ostría Gutiérrez, Alberto. *Una obra y un destino: La política internacional de Bolivia después de la guerra del Chaco*. Buenos Aires: Ayacucho, 1946.

Palacios, Alfredo L. *Petróleo, monopolios y latifundios*. 2d ed. Buenos Aires: Kraft, 1957.

Pavlovsky, Aaron. *La cuestión agraria*. Buenos Aires: Peuser, 1913.

Pearton, Maurice. *Oil and the Rumanian State*. Oxford: Clarendon Press, 1971.

Pegazzano, Manuel C. "Explotación mixta de los yacimientos de petróleo de Comodoro Rivadavia," in Universidad Nacional de Buenos Aires, Facultad de Ciencias Económicas, *Investigaciones de Seminario*, 2 (1921), pp. 319–38.

Penna Marinho, Ilmar. *Petróleo: Soberanía & desenvolvimento*. Rio de Janeiro: Bloch, 1970.

Penrose, Edith T. *The Large International Firm in Developing Countries: The International Petroleum Industry*. Cambridge, Mass.: MIT Press, 1968.

Pérez Prins, Ezequiel. "La refinería de petróleo de la ANCAP." *Boletín del Instituto Sudamericano del Petróleo* (Montevideo), 1 (Apr. 1943), pp. 15–36.

Perón, Juan Domingo. *La fuerza es el derecho de las bestias*. Bogotá: Villegas, 1956.

Peters, H. E. *The Foreign Debt of the Argentine Republic*. Baltimore, Md.: Johns Hopkins, 1934.

El petróleo del norte argentino: Comentarios del diario "El Intransigente" de la ciudad de Salta, con un prólogo del General Enrique Mosconi. Salta, Argentina: C. Velarde, 1928.

El petróleo. Su importancia en la vida de las naciones. Necesidad de una legislación nacional sobre petróleo. Conferencias en pro de su sanción. Buenos Aires: Luis Bernard, 1927.

The Petroleum Almanac. New York: National Industrial Conference Board, 1946.

Phelps, Dudley M. *The Migration of Industry to South America*. New York: McGraw-Hill, 1936.

————. "Petroleum Regulation in Temperate South America," *American Economic Review*, 29 (Mar. 1939), pp. 48–59.

Phelps, Vernon L. *The International Economic Position of Argentina*. Philadelphia: Univ. of Pennsylvania Press, 1938.

Pike, Frederick B. "Making the Hispanic World Safe From Democracy: Spanish Liberals and *Hispanismo*," *The Review of Politics*, 33 (Jul. 1971), pp. 307–22.

Pinelo, Adalberto J. *The Multinational Corporation as a Force in Latin American Politics: A Case Study of the International Petroleum Company in Peru*. New York: Praeger, 1973.

Platt, D.C.M. *Latin America and British Trade, 1806–1914*. London: Adam and Charles Black, 1972.

Potash, Robert A. *The Army and Politics in Argentina, 1928–1945: Yrigoyen to Perón*. Stanford, Calif.: Stanford Univ. Press, 1969.

Powell, J. Richard. *The Mexican Petroleum Industry, 1938–1950*. Berkeley: Univ. of California Press, 1956.

Puga Vega, Mariano. *El petróleo chileno*. Santiago: Andrés Bello, 1964.

Quesada, Ernesto. *El "peligro alemán" en Sud América*. Buenos Aires: Selín Suárez, 1915.

Randall, Stephen J. "The International Corporation and American Foreign Policy: The United States and Colombian Petroleum, 1920–1940," *Canadian Journal of History*, 9 (Aug. 1974), pp. 179–96.

Repetto, Nicolás. *Mi paso por la política: De Roca a Yrigoyen*. Buenos Aires: Santiago Rueda, 1956.

Reyes, Cipriano. *Yo hice el 17 de octubre*. Buenos Aires: GS Editorial, 1973.

Rippy, J. Fred. "The United States and Colombian Oil," *Foreign Policy Association Information Service*, 5 (Apr. 3, 1929), pp. 20–35.

Robertson, Malcolm. "The Economic Relations Between Great Britain and the Argentine Republic," *International Affairs*, 9 (Mar. 1930), pp. 222–31.

Rock, David. "Machine Politics in Buenos Aires and the Argentine Radical Party, 1912–1930," *Journal of Latin American Studies*, 4 (Nov. 1972), pp. 233–56.

————. *Politics in Argentina, 1890–1930: The Rise and Fall of Radicalism*. London: Cambridge Univ. Press, 1975.

————. "Radical Populism and the Conservative Elite, 1912–1930," in Rock, ed., *Argentina in the Twentieth Century*, pp. 66–87.

————. "The Survival and Restoration of Peronism," in Rock, ed., *Argentina in the Twentieth Century*, pp. 179–221.

Rock, David, ed. *Argentina in the Twentieth Century*. Pittsburgh, Pa.: Univ. of Pittsburgh Press, 1975.

Rodríguez, Celso. "Regionalism, Populism, and Federalism in Argentina, 1916–1930." Ph.D. diss., Univ. of Mass., 1974.

Roggi, Luis O., ed. *Argentina: Petróleo y soberanía, 1955–1964*. Cuernavaca, Mexico: CIDOC, 1968.

Rout, Leslie B. *Politics of the Chaco Peace Conference, 1935–1939*. Austin: Univ. of Texas Press, 1970.

Ruiz Guiñazú, Enrique. *Las fuerzas perdidas en la economía nacional*. Buenos Aires: "El Ateneo" 1917.

Rumbo, Eduardo I. *Petróleo y vasallaje: Carne de vaca y carnero contra carbón más petróleo*. Buenos Aires: Hechos e Ideas, 1957.

Russell, Ronald S. *Imperial Preference: Its Development and Effects*. London: Empire Economic Union, 1947.

Rutledge, Ian. "Plantations and Peasants in Northern Argentina: The Sugar Cane Industry of Salta and Jujuy, 1930–1943," in Rock, ed., *Argentina in the Twentieth Century*, pp. 88–113.

Sábato, Arturo. *Petróleo: Liberación o dependencia*. Buenos Aires: Macacha Güemes, 1974.

Sáenz Peña, Roque. *Escritos y discursos*. 2 vols. Buenos Aires: Peuser, 1935.

Salera, Virgil. *Exchange Control and the Argentine Market*. New York: Columbia Univ. Press, 1941.

Sampson, Anthony. *The Seven Sisters: The Great Oil Companies and the World They Made*. New York: Viking Press, 1975.

Sánchez Sorondo, Matías G. *La palabra de un patriota sobre el problema de la legislación del petróleo*. Buenos Aires: V. Domínguez, 1927.

Sánchez Viamonte, Carlos. *El último caudillo*. 2d ed. Buenos Aires: Devenir, 1956.

Sargent, A. J. *Coal in International Trade*. London: P. S. King and Son, 1926.

Scalabrini Ortiz, Raúl. *Política británica en el Río de la Plata*. Buenos Aires: Reconquista, 1940.

Scobie, James R. *Buenos Aires: Plaza to Suburb, 1870–1914*. New York: Oxford Univ. Press, 1974.

Selser, Gregorio. *Argentina a precio de costo: El gobierno de Frondizi*. Buenos Aires: Iguazú, 1965.

Serghiesco, Traian T. *Líneas generales sobre los Yacimientos Petrolíferos Fiscales desde el punto de vista técnico-geológico y económico*. Buenos Aires: La Aurora, 1930.

Silenzi de Stagni, Adolfo. *El petróleo argentino*. 2d ed. Buenos Aires: Colección Problemas Nacionales, 1955.

———. "Prólogo," in Centro de Estudios General Mosconi, *Los tratantes de petróleo*, pp. 7–13.

Smith, L. Brewster, Harry T. Collings, and Elizabeth Murphy. *The Economic Position of Argentina During the War*. Washington, D.C.: Dep't of Commerce, Bureau of Foreign and Domestic Commerce, 1920.

Smith, Peter H. *Argentina and the Failure of Democracy: Conflict Among Political Elites, 1904–1955*. Madison: Univ. of Wisconsin Press, 1974.

———. *Politics and Beef in Argentina: Patterns of Conflict and Change*. New York: Columbia Univ. Press, 1969.

———. "The Social Base of Peronism." *Hispanic American Historical Review*, 52 (Feb. 1972), pp. 55–73.

Smith, Peter Seaborn. *Oil and Politics in Modern Brazil*. Toronto: Macmillan, 1976.

Smith, Robert Freeman. *The United States and Revolutionary Nationalism in Mexico, 1916–1932*. Chicago: Univ. of Chicago Press, 1972.

Smith, Robert Richard. "Radicalism in the Province of San Juan: The Saga of

Federico Cantoni (1916–1934)." Ph.D. diss., Univ. of Calif., Los Angeles, 1970.

Snow, Peter G. *Argentine Radicalism: The History and Doctrine of the Radical Civic Union*. Iowa City: Univ. of Iowa Press, 1965.

Sociedad Anónima "Iuyamtorg." *Exposición-Feria, 1928, Buenos Aires*. Buenos Aires: L. J. Rosso, 1928.

Solberg, Carl E. "Rural Unrest and Agrarian Policy in Argentina, 1912–1930," *Journal of Inter-American Studies and World Affairs*, 13 (Jan. 1971), pp. 18–52.

———. "The Tariff and Politics in Argentina, 1916–1930." *Hispanic American Historical Review*, 53 (May 1973), pp. 252–74.

Sommariva, J. O. *La república federal, 1912–1936*. La Plata, Argentina: "Olivieri y Domínguez," 1955.

Sommi, Luis V. *Hipólito Yrigoyen: Su época y su vida*. Buenos Aires: Monteagudo, 1947.

Spinelli, Joaquín. "En pleno juicio final," *Petróleo y Minas*, 7 (Aug. 1, 1927), pp. 6–7.

Stephens, Tomás E. "La influencia de la Argentina en la lucha futura," *Petróleo y Minas*, 2 (Apr. 15, 1922), p. 9.

Sweet, Dana Royden. "A History of United States-Argentine Commercial Relations, 1918–1933: A Study of Comparative Farm Economies." Ph.D. diss., Syracuse Univ., 1971.

Tanzer, Michael. *The Political Economy of Oil and the Underdeveloped Countries*. Boston: Beacon Press, 1969.

Tornquist, Ernesto & Cía., Ltda. *El desarrollo económico de la República Argentina en los últimos cincuenta años*. Buenos Aires: Tornquist 1920.

Torres, Francisco S. "Aviación: fomento e industria," *Revista Militar*, 24 (Jan. 1924), pp. 69–75.

———. "Aviones metálicos, aviones junkers," *Revista Militar*, 25 (Apr. 1925), pp. 555–88.

Troncoso, Oscar A. *Los nacionalistas argentinos: Antecedentes y trayectoria*. Buenos Aires: S.A.G.A., 1957.

Tugendhat, Christopher. *Oil: The Biggest Business*. London: Eyre & Spottiswoode, 1968.

Tugwell, Franklin. *The Politics of Oil in Venezuela*. Stanford, Calif.: Stanford Univ. Press, 1975.

Tulchin, Joseph S. *The Aftermath of War: World War I and U.S. Policy Towards Latin America*. New York: New York Univ. Press, 1971.

———. "The Argentine Economy During the First World War," *The Review of the River Plate*, 147 (Jun. 19, 1970), pp. 901–3; 147 (Jun. 30, 1970), pp. 44–46.

———. "Foreign Policy," in Falcoff and Dokart, eds., *Prologue to Perón: Argentina in Depression and War, 1930–1943*, pp. 83–109.

Ugarte, Manuel. "El ocaso socialista y la guerra europea," *Boletín de la Unión Industrial Argentina*, 30 (June 15, 1916), pp. 2–8.

[Unión Industrial Argentina.] *Estado actual de la industria del petróleo en la República*

Argentina y su posible desenvolvimiento en lo futuro. Nota presentada a la Honorable Cámara de Diputados de la Nación por el gremio de compañías industriales de petróleo adheridas a la Unión Industrial Argentina. Buenos Aires: "Coni," 1927.

United States, Official and Semiofficial Publications:
[1] Department of State. *Foreign Relations of the United States: Diplomatic Papers, 1936. V: The American Republics.* Washington, D.C.: Dep't of State, 1954.
[2] ————. *Foreign Relations of the United States: Diplomatic Papers, 1945. IX: The American Republics.* Washington D.C.: Dep't of State, 1969.
[3] ————. *Records of the Department of State. Decimal File, 1910–29: Argentina, Internal Affairs.* File M-514. National Archives, Washington, D.C.

Van Niekerk, A. E. *Populism and Political Development in Latin America.* Rotterdam: Rotterdam Univ. Press, 1974.

Velarde, Carlos E. *Las minas de petróleo en la legislación argentina.* Buenos Aires: "Coni," 1922.

Vicat, Luis E. "Combustibles y defensa nacional," *Revista Militar,* 23 (Sep. 1923), pp. 347–53; 23 (Oct. 1923), pp. 471–79; 23 (Nov. 1923), pp. 607–10; 23 (Dec. 1923), pp. 769–92; and 24 (Jan. 1924), pp. 19–22.

————. "Ideas sueltas que creo útiles al progreso del norte argentino," in Villafañe, *El atraso del interior,* pp. 177–223.

————. "Necesidad de una metalúrgica propia como elemento indispensable para asegurar la defensa nacional," *Revista Militar,* 26 (Aug. 1925), pp. 123–39.

Villafañe, Benjamín. *El atraso del interior: Documentos oficiales del gobierno de Jujuy pidiendo amparo para las industrias del norte.* Jujuy, Argentina: B. Buttazzori, 1926.

————. *Degenerados. Tiempos en que la mentira y el robo engendrán apóstoles.* Buenos Aires: no pub., 1928.

————. "Gravedad que encierra la cuestión del petróleo," in Sánchez Sorondo, *La palabra de un patriota,* pp. 3–56.

————. *La región de las parias.* Buenos Aires: Cabaut, 1934.

————. *El yrigoyenismo: No es un partido político. Es una enfermedad nacional y un peligro público.* Jujuy, Argentina: Talleres Gráficos del Estado, 1927.

Villafañe, Benjamín, ed. *El petróleo y la constitutión nacional.* Jujuy, Argentina. Talleres Gráficos del Estado, 1926.

Villanueva, Javier. "Economic Development," in Falcoff and Dolkart, eds., *Prologue to Perón,* pp. 57–82.

Villar Araujo, Carlos. "Informe sobre el petróleo en la Argentina," *Crisis,* 24 (Apr. 1975), pp. 11–21; 25 (May 1975), pp. 3–11.

Volski, Victor. *América Latina, petróleo e independencia.* Buenos Aires: Cartago, 1966.

Votaw, Dow. *The Six-Legged Dog: Mattei and ENI—A Study in Power.* Berkeley: Univ. of California Press, 1964.

Walter, Richard J. *The Socialist Party of Argentina, 1890–1930.* Austin: Univ. of Texas Press, 1977.

————. *Student Politics in Argentina: The University Reform and its Effects, 1918–1964*. New York: Basic Books, 1968.

Whitaker, Arthur P. *Argentine Upheaval: Perón's Fall and the New Regime*. New York: Praeger, 1956.

Wilkins, Mira. *The Maturing of Multinational Enterprise: American Business Abroad from 1914 to 1970*. Cambridge, Mass.: Harvard Univ. Press, 1974.

————. "Multinational Oil Companies in South America in the 1920s: Argentina, Bolivia, Brazil, Chile, Colombia, Ecuador, and Peru," *Business History Review*, 48 (Fall 1974), pp. 414–46.

Williams, John H. "Argentine Foreign Exchange and Trade Since the Armistice," *Review of Economics and Statistics*, 3 (Mar. 25, 1921), pp. 47–57.

Wirth, John D. *The Politics of Brazilian Development, 1930–1954*. Stanford, Calif.: Stanford Univ. Press, 1970.

Wood, Bryce. *The Making of the Good Neighbor Policy*. New York: Columbia University Press, 1961.

World Petroleum Report: An Annual Review of International Oil Operations. Dallas, Texas: M. Palmer, 1970, 1971, 1973.

Wright, Winthrop R. *British-owned Railways in Argentina: Their Effect on the Growth of Economic Nationalism, 1854–1948*. Austin: Univ. of Texas Press, 1974.

Yrigoyen, Hipólito. *Mi vida y mi doctrina*. Buenos Aires: Raigal, 1957.

————. *Pueblo y gobierno*. 12 vols. Buenos Aires: Raigal, 1954–56.

Zinser, James E. "Alternative Means of Meeting Argentina's Petroleum Requirements," in Mikesell, *Foreign Investment in the Petroleum and Mineral Industries*, pp. 189–215.

————. "Alternative Means of Satisfying Argentine Petroleum Demand: Importation, Government Production, or Foreign Private Contractual Production: A Comparative Analysis and a Recommended Petroleum Policy." Ph.D. diss., Univ. of Oregon, 1967.

Zuvekas, Clarence. "Argentine Economic Policy, 1958–1962: The Frondizi Government's Development Plan," *Inter-American Economic Affairs*, 22 (Summer 1968), pp. 45–73.

Index